LIFE OF THE PAST

An Introduction to Paleontology

BY GEORGE GAYLORD SIMPSON, 1902-

New Haven and London: YALE UNIVERSITY PRESS

First published April 1953
Issued as a Yale Paperbound June 1961
Ninth printing, November 1967

Printed in the United States of America by
Vail-Ballou Press, Inc., Binghamton, N.Y.,
and reprinted by The Carl Purington Rollins Printing-Office
of the Yale University Press,
New Haven, Connecticut

Library of Congress catalog card number: 52-12078

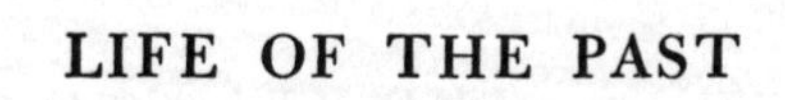

LIFE OF THE PAST

By the same author

THE MEANING OF EVOLUTION

To my daughters:

HELEN
GAY
JOAN
BETTY

Preface

TO THOSE who follow that science, nothing is more lively than the study of the dead remains of ancient life. Fascinating in itself, paleontology also impinges on much else that is interesting, important, and useful.

It has been difficult for the general reader and the student alike to find a comprehensible but not condescending account of the most widely interesting aspects of paleontology. Most books on this subject stress the description of particular sorts of fossils. They do not adequately explain what the paleontologist is really getting at, and why. There is, moreover, a gap between popular books on prehistoric animals that are either juvenile or journalistic and textbooks that are heavily concerned with professional training in technique and terminology.

There should be available a nontechnical discussion of the whole scope and significance of paleontology as a science, concerned with the principles and interpretation of the history of life and not only with the identification of objects called fossils. That, in brief, was the motive for writing this book. The reader is the best judge whether the need has been sufficiently filled.

Acknowledgments may also be brief. A book covering so wide a field is a distillation of the work of others so numerous that they cannot be listed. I am more specifically indebted to my wife, Dr. Anne Roe, for encouragement and for critical reading of each chapter to the great improvement of the book's readability. I am also pleasantly indebted to Miss Roberta Yerkes and the others at the Yale University Press who welcomed this book, worked with it, and now present it to you.

Mr. David B. Kitts has done most of the work on the index.

The illustrations, which (it is hoped) may make up in clarity what they lack in artistry, were all newly drawn by me. Data have come from many sources. When possible a source is credited, but no one else is responsible for either illustrations or text in just their present form.

G. G. SIMPSON

Los Pinavetes, La Jara, New Mexico
August, 1952

Contents

Illustrations

1. *Prelude: A Walk Through Time*

Let us take a long walk in northwestern New Mexico. We may start from the rim of Chaco Canyon, where we can look down the vertical cliff and see a ruined pueblo, an ancient stone apartment house with 800 or more rooms. This one-house town was already old in the 12th century but was flourishing then. Ahead of us to the northeast stretches undulating country, most of it pungent and gray-green with low sagebrush. Here and there along rocky crests are dark lines of bushy tree junipers and twisted pinyon-nut pines. On occasional rises along the sides of the broad, dry watercourses, the sandy arroyos of the Southwest, are areas of badlands. These are patches, from a few yards to many miles in extent, where the earth's crust is laid bare and has been carved by wind and rain into fantastic and austerely beautiful forms—wildernesses of twisted gorges, banded clay slopes, and weird sandstone-capped pillars.

We will seem to be solitary on our walk, striding across the vast, empty earth beneath the even vaster, incredibly blue, empty sky; but we will not be alone. Occasionally an antelope ground squirrel will dart across our trail or a coyote will slink away wraithlike at our approach. Always we will be watched by the alert dark eyes of Navajo Indians, crouching in the brush to guard their scattered flocks and aloofly curious about the actions of the white strangers.

A few miles from the canyon rim is a broad badlands area eroded in soft white and brown sandstone and gray clays with black lines that are the exposures of thin beds of coal. That coal, composed of ancient vegetation, reveals an exuberance of former plant life in strange contrast with the semidesert of today. A few minutes of prospecting along the clays reveal something stranger still: great bones much larger than those of any living American animal are being washed out by storms from their burial places in the hard clay. They are the remains of dinosaurs that swarmed here when the country was a lush lowland.

Another few miles and we come to a ledge of yellow sandstone in which are imbedded many large logs, fallen tree trunks feet in diameter and yards long. They look like fresh wood, inviting to the ax, but an ax swung against them would shatter its edge and strike sparks, for this wood is now silica, hard as rock crystal of which it is, indeed, a form. More extensive search shows that this sandstone, too, contains scattered bones of giant reptiles, dinosaurs known by such names as *Kritosaurus, Pentaceratops,* and *Alamosaurus.*

At the top of this sandstone there is a sharp line of division between it and a series of clays, banded in tones of gray with an occasional wine-red layer and lenses of white sand. Mark that line well, for it is the geological trace of one of the most dramatic events in the history of life. If we follow one of the wine-red bands a few feet above the division line and if we have the patience to keep our eyes glued to the ground for hour after hour and to crawl on hands and knees over the rough surface, eventually we will find ancient bones and teeth again. But what a difference! Here are no dinosaur bones. Most of these remains come from animals no larger than squirrels, and the largest of them might have to stretch to look a sheep in the face. If we study them closely we will find, too, that these bones and teeth are anatomically very different from those of dinosaurs. They are not remains of cold-blooded reptiles but of mammals, creatures warm-blooded like ourselves but smaller, vastly more ancient and more primitive than any human being.

We continue walking, still northeastward, and continue searching diligently for such remains of life as may be washing out from burial places in the successive clays and sandstones. By now, actually, our walk will have lasted for days, perhaps for weeks. We are covering only about thirty miles in an air line, but the ancient bones and teeth for which we are searching in successive layers are not abundant. At each level we must spend hours and often days to find any well enough preserved to tell us a clear story. With persistence we do find them, and we notice that their character is changing as we go along. No more dinosaurs are found, but the remains of mammals become more varied, some of the teeth become progressively more complicated in pattern, and larger bones do begin to appear.

We may stop finally among high mesas along the slopes of which are badlands magnificently banded in shades of red, lavender, yellow and gray. Here we are almost sure to find, within a day or two, bones of an animal not unlike some we saw earlier in our walk but considerably larger and different in anatomical detail. *Coryphodon,* the giant of its time, could be compared for size with a particularly squat cow. Equally or more probable is the discovery of smaller remains quite unlike anything we have seen in the lower country behind us and of unusual interest to us. These belong to eohippus, the correct scientific name of which unfortunately happens to be *Hyracotherium*. A long sequence of discoveries elsewhere has shown that eohippus is, indeed, the "dawn horse," earliest known ancestor of Dobbin. With more time and luck we may even find a bit of the jaw of one of several sorts of animals smaller still and more interesting still: pre-monkeys, our own relatives some 60 million years removed.

This has been a walk through time. We have seen and touched a long segment of truly ancient history. Above the crust of the earth there have been tokens of shorter, human history: the pueblo ruin, the later Navajo invaders, and the latest invaders, represented by ourselves. This history is one of the outcomes of the longer history, but its few centuries pale to insignificance in comparison with the span we have followed in the exposed crust of the earth.

That span covered some 20 million years. It began toward the end of the Age of Reptiles while dinosaurs still ruled, perhaps 80 million years ago, and ended as the Age of Mammals was getting well under way, more or rather less than 60 million years ago. (I say "some," "perhaps," and "more or less" because these dates in years are not accurately determined, as will be explained later.) The line of division between the yellowish, log-bearing sandstone, which geologists call the Ojo Alamo formation, and the overlying clays, the Nacimiento formation, was the line between the Age of Reptiles, the earth's Medieval Age or Mesozoic Era, and the Age of Mammals, the earth's Modern Age or Cenozoic Era. (See Fig. 1.)

At that line the last dinosaurs became extinct and the mammals, warm-blooded, furry, milk-giving, took over. At that time the mammals were still small, primitive, and not particularly varied, even though they already had a long history behind them. As we continued our walk we were witnesses to the expansion of the mammals, their progressive diversification, and the appearance of increasingly modern types.

The beginning of knowledge of the history of life comes from many, many walks like that by innumerable searchers in all parts of the world. The accumulated remains, with careful notes as to their places and sequence in the earth's crust, are sent to labora-

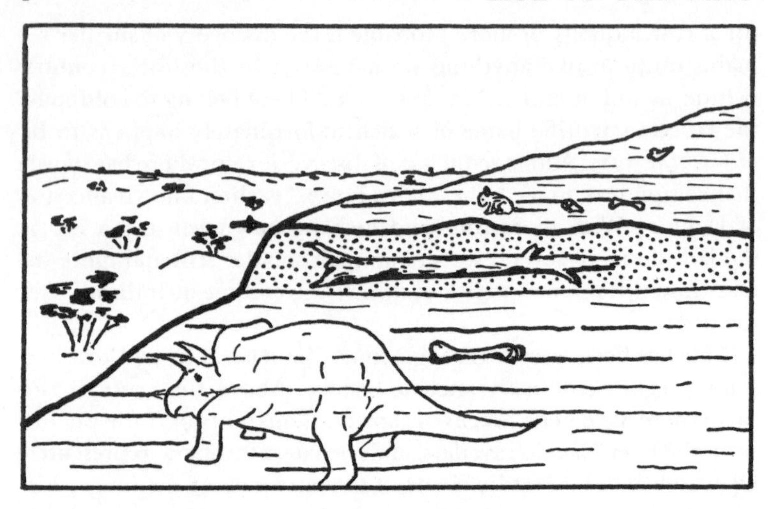

FIG. 1. Part of the rock succession in northwestern New Mexico. The lower rocks contain remains of dinosaurs, one of which is shown in ghostly restoration. The arrow points to the line between the Age of Reptiles and the Age of Mammals. Above this there are no more dinosaurs, but remains of small warm-blooded animals occur.

tories for preparation and study. They are identified and named. Their places in the past economy of nature are considered. Their relationships and lines of descent through the ages are determined. Little by little there emerges an ever clearer picture of what has happened in the history of life. More important still, we begin also to see just how life has changed and to be able to judge why it has changed and why it is now what it is.

We are, ourselves, products of that history and we are its heirs. We cannot understand our own place in the universe or wisely guide our own affairs without knowledge of the processes that produced us and that still affect us and all the life around us. That fact alone makes knowledge of the principles of the history of life imperative for modern man, even if the history itself were not of intrinsic interest—and few who look into it fail to feel its fascination.

Although it is not the purpose of this book to trace the history in detail, its broader features will be outlined. Of equal or greater value and more pertinent to the present purpose are such topics as how the history is deciphered, what ancient animals did and how we know, how they lived together and under what circumstances, their spread over the earth and the past geography of the earth, the ways of organic change, and especially the evidence of past life on the nature of life, the reasons for its evolution, and the meaning of all this to us today. Those are the main subjects of the following chapters.

2. *The Remains of Ancient Life*

IF YOU walk in New Mexico and pick up what is plainly a bone weathering out from ancient clays, or along the slopes of the Catskills and find what is plainly a sea shell embedded in the rocks, you can say at once, "That is a fossil; it is a remnant of an animal that lived here long ago." The fact now seems quite obvious. It was far from obvious to many of our ancestors and was only generally accepted as a fact after centuries of learned argument. The long history is a good example of how painful and groping is man's rise to knowledge and how faith, dogma, and authority can make us blind to the plain evidence of our senses.

As early as the 6th century before Christ the Greeks knew in a general way what fossils were and what they mean, but as late as the 18th century of our era, some 2200 years later, men of science were still gravely arguing the point. A long list of Greek philosophers and historians, among them Xenophanes, Xanthos, and Herodotus, noticed that sea shells may be found buried far inland and concluded that the sea had once stood where the shells were found. Bones of mammoths were also known to the ancients. They recognized these as bones but usually ascribed them to gigantic men —an interpretation still generally accepted in the 18th century and occasionally thereafter.

During the Dark and Middle Ages the ancient lore about fossils, rudimentary as it was, was largely forgotten. Among those who paid any attention to fossils at all the most popular theories were that they had been engendered in the rocks by a sort of "formative force," or that they had grown in the ground from germs fallen from the stars, or in some cases had fallen fully formed from the heavens. No clear distinction was made between what we now call fossils, that is, the actual remains or traces of ancient living things, and such things as crystals, old stone axes, or rocks that happen to have the shape of an ear or some other organ or animal. Even in the 16th century, when Agricola coined the term "fossil" (from Latin *fossilis,* "dug up"), it meant any curious object found buried. Fig. 2 shows some of the things that Gesner called fossils in 1565.

Only gradually thereafter did the word come to be confined to traces of life, the only usage now current.

While most scholars were arguing fluently and almost entirely erroneously as to the nature of what we now call fossils, an opposite error was becoming widely accepted. In the 13th century Albertus Magnus was probably echoing general opinion when he said that "Whole animals can be petrified. The constituents of the body of the animals are modified. Earth mixes with water and mineral matter changes the whole thing to stone while preserving the form of the animal." We will see below that something rather like this can happen but that it is extremely rare in animal fossils. The objects to which Albertus was referring were almost certainly not petrified animals but merely stones with an accidental resemblance, like the "ear stones" mentioned above. This idea dies hard, as all mistakes seem to. Museums are still being bombarded with "petrifactions" the only connection of which with a fossil animal is in the eye of the beholder. Only the other day I was offered for sale at a large price "the petrified leg of a woman." I was called a liar and a cheat when I explained that it was only a piece of volcanic rock with an accidental (and very slight) resemblance to the vision in the mind of its owner.

FIG. 2. Some of Gesner's fossils (1565). A, a quartz crystal. B, a belemnite, part of an extinct relative of the cuttlefishes. C, a prehistoric stone ax. Only B would be called a fossil today.

A few of the learned did retain the ancient idea that fossil shells were shells in fact and that where they occur the sea has been. To be learned then was to be a cleric, and it was inevitable that sooner or later someone would have the wonderful idea that fossils are witnesses to the Biblical deluge. Ristoro d'Arezzo, an Italian monk of the 13th century, is sometimes credited with this inspiration.

Whether he was first or not, he did express the idea in a work on the "Composition of the World" in A.D. 1282. This error, too, has never died out altogether. It is still taught as, literally, gospel truth in some church-controlled schools in the United States. In the intellectual atmosphere of the Middle Ages, which these anachronistic schools still breathe, it is not surprising that this soon became the predominant theory as to fossils.

Leonardo da Vinci in the late 15th century was a precursor of modern geology as of so much else that is modern. He rejected alike the views that fossils are mere "plays of nature" and that they are testimony to the deluge. He said, "The mountains where there are shells were formerly shores beaten by waves, and since then they have been elevated to the heights we see today," and he produced logical, keenly observed evidence for this correct opinion.

Thereafter interest in fossils was more intense and continuous than before, but for two centuries this meant that agreement as to their significance was even less general and disputes were even more bitter. If space permitted these centuries would yield a noble roster of honest observation and logical deduction, an infamous roster of educated bigotry, and a ridiculous roster of superstition and surmise. High on the first list belongs, for instance, Bernard Palissy, a French pottery maker and self-educated naturalist. He collected fossils and concluded that they were exactly what we now know them to be. He even correctly identified the species of some of them in a thoroughly scientific way, probably the first time that this was ever done. Finally he went to Paris and gave a series of lectures attacking the dignitaries of the Sorbonne, whose opposition won them places well up among the educated bigots. Palissy was right, but the Sorbonne won: Palissy died in the Bastille in 1590.

Nearly a century after Palissy's ill-omened debate a Dane, Nicolaus Steno, living in Florence, published a book dated 1669 in which he not only recognized the true nature of fossils but also pointed out the successional nature of rock strata. The works of Palissy and of Steno, between them, contain all the essentials on which a true science of paleontology, of fossils and the history of life, could have been and was, indeed, later to be based. Yet it was again well over a century after Steno until it was definitely established and generally agreed that there is a succession of different fossils in the rocks and that this succession is the record of the history of life. Of course others contributed greatly, before and after, but this final triumph was largely due to William Smith (1769–1839) in England, G. B. Brocchi (1772–1826) in Italy, and Alexandre Brongniart (1770–1847) in France.

Even so brief an account must not be ended with the impression that history moves steadily forward, that knowledge and understanding progressed from da Vinci to Palissy to Steno to Smith and onward to us. A few brief indications to the contrary: In 1699 Edward Lwhyd published what was for the time a magnificent atlas of fossils and explained to the reader, as medieval monks had said long before, that these were engendered in the rocks from seeds carried by wind and water. In 1695 John Woodward used his fine collection of fossils as a basis for "an Account of the Universal Deluge." J. J. Scheuchzer, Swiss, was so delighted with Woodward's book that he followed it by several of his own, including one that proved from fossil evidence that the deluge occurred in the month of May. (This conflicted with the previous demonstration by an English astronomer, William Whiston, that the flood occurred on a Wednesday, November 28.) In 1746 Voltaire, a less successful universal genius than da Vinci, wrote, "Is it really sure that the soil of the earth cannot give birth to fossils? A tree has not produced the agate which perfectly portrays a tree. Similarly, the sea may not have produced these fossil shells which look like the homes of little marine animals." And in 1952 there is a purported textbook of geology on the market that would have seemed laughably antiquated to Voltaire.

In spite of the fact that old errors never die, it has been general knowledge among competent students for about two centuries now that fossils are the remains of ancient organisms, that they can be identified and classified by anatomical comparison, that they occur in sequence in certain rocks of the earth's crust, and that they form a historical record. There has been much more recent dispute as to the meaning of that record, but that is a different point to which we will come later. Now let us consider what sort of things fossils are.

How wonderful it is that organisms so ancient can be preserved for us to see, handle, and study! Plants and animals die around us today, and we usually find no recognizable trace of them a few years, a few months, sometimes even a few hours later. Yet some fossil plants are well preserved after more than a thousand million years, and the evanescent delicacy of a jellyfish has lasted as much as 500 million years.

Preservation of an animal entire, just as it died, is the rarest of accidents and with a few exceptions has persisted only for organisms so recent as hardly to merit designation as fossils. They are called fossils principally because they are of species that became extinct in prehistoric times. Any remains really ancient, from, say, 100 thousand years upward, are agreed to be fossils no matter how they may be preserved. Younger fossils, on into the dawn of history, intergrade with recent remains and prevent hard and fast definition. There is no exact point when an animal becomes a fossil. Certainly this is not when it petrifies—most fossils never do really petrify.

Extinct animals preserved whole, or nearly so, are then a sort of subfossil. They include the famous mammoths found in Siberia, frozen whole and preserved in cold storage for some thousands or perhaps at most tens of thousands of years. In Alaska, too, considerable parts of mammoths and of some other animals have been preserved in frozen muck. Mammoth hair is sometimes common enough there to be a nuisance in the gold diggings. In Starunia, Galicia, a young mammoth and a young woolly rhinoceros were found in a deposit of waxy hydrocarbon which had preserved them nearly whole. In particularly dry caves in southwestern United States and in Patagonia were found ground sloths perhaps a few thousand years old, not whole, to be sure, but with large parts of desiccated hide and hair, tendons, and piles of excrement preserved in addition to the bones.

Interesting as they are, such finds are so rare that they are more curiosities than usual objects of paleontological study. More common and of more scientific importance are the occurrences of parts of plants and of partial or whole animals, mostly insects, in amber. Amber is the hardened, chemically altered resin of ancient trees. When the resin was soft many insects and a few other organisms were trapped and sealed in it, embedded in a tomb of transparent, antiseptic, natural plastic. The famous Baltic amber used in jewelry from ancient times is particularly rich in beautifully preserved whole insects. Because the amber where found is not in its original deposit but has been washed in from somewhere else, its exact age is unknown. It may be on the order of 50 million years old, and amber with insects of possibly still greater age is known from other regions.

There are a few other antiseptic burial places where soft tissues may be preserved with little change. Noteworthy among these are bogs. Soft coals formed in ancient bogs may be tens of millions of years old, as in Victoria, Australia, and contain wood that is somewhat darkened but otherwise so unaltered that it can easily be cut with a saw and planed. In Germany a similar deposit has produced remains of animals badly flattened and distorted but with some tissues so fresh that details of soft cells can be seen under the microscope. The bottom waters of some lakes and arms of the sea can also become antiseptic with accumulation of chemicals so that decay does not occur there. To this circumstance we owe such extraordinary finds as fishes, also flattened and blackened, but so preserved that after nearly 300 million years the muscle fibers and their cross-striations are still visible with a microscope.

Practically speaking, truly petrified animals with the whole body, soft parts and all, turned to stone do not occur, in spite of Albertus Magnus and some enthusiastic but ill-informed moderns. Perhaps the nearest thing to such preservation is that of whole bodies of frogs and a few other animals replaced by phosphate minerals near Quercy in France. The occurrence is altogether exceptional, indeed unique.

The preservation of wholly soft animals without skeletons and of soft parts of animals with skeletons is not very rare, but it commonly occurs in two ways only: as thin films of carbon and as plastic impressions on the rocks. All soft tissues contain carbon in complex compounds. In the processes of partial decay and burial, with increasing pressure as fossil-bearing rocks accumulate, these compounds may break down in such a way that other volatile or soluble materials escape but a thin film of black carbon is left. That is the way the 500-million-year jellyfish was preserved, and with it a fascinating array of other soft-bodied creatures. Most of these were collected in British Columbia, Canada, and described by the late American paleontologist C. D. Walcott. It is also the explanation of the preserved body outlines of the famous ichthyosaurs from Holzmaden, Germany, but even there this is a rarity. Most of the ichthyosaurs are preserved as skeletons only. Preservation of fossil

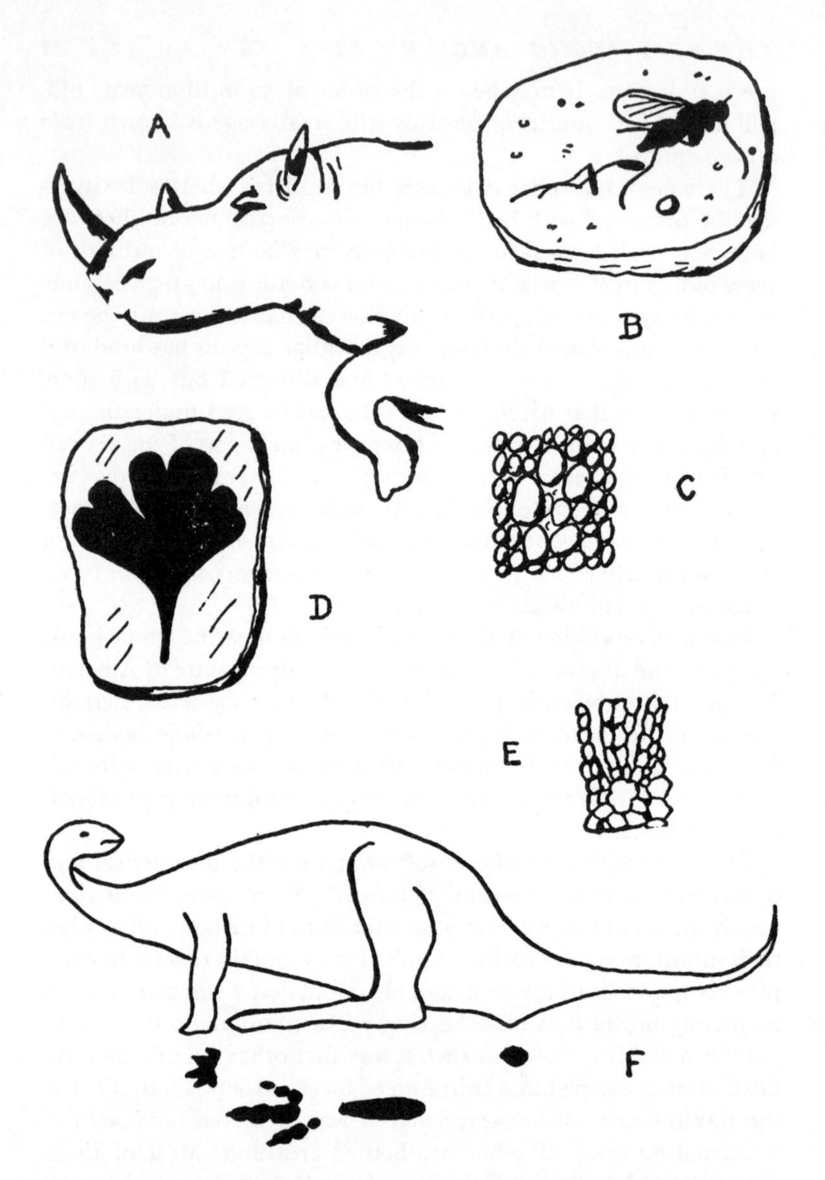

FIG. 3. Some ways in which fossils are preserved. A, front end of an extinct rhinoceros, preserved entire in mineral wax (after a photograph by Niezabitovski). B, a fly in amber (after a photograph by Bachofen-Echt). C, impression of dinosaur (hadrosaur) skin in sandstone. D, leaf (ginkgo) preserved as a film of carbon. E, petrified wood, greatly en-

leaves as carbon films is very common, indeed usual, but there is usually a plastic impression also.

The rocks in which fossils occur were almost always soft sediments, muds, silts, and sands, when the remains were buried. Even the most delicate tissues may make an imprint on such sediments, and if another layer is deposited without destroying the imprint it may be preserved as the rock hardens and may last indefinitely. Fossil jellyfish have been recorded in this way, too. Or the body may decay after burial and leave a cavity, which may remain as a hollow in the rock or be filled by sediment or mineral later on. One of the most remarkable instances was recently found in Oregon. There an ancient rhinoceros, apparently already dead but with the skin still intact, was overwhelmed by a lava flow that quickly hardened around its body. What remains now is a cavity reproducing the shape of the bloated animal, and inside the cavity were a few fragments of burned bones.

The famous dinosaur "mummies" are not in the usual sense mummies. After the animals died the skin dried over the skeleton and the whole thing was then buried in sand. The actual skin and all the soft internal organs have long since disappeared, but clear impressions of skin are preserved on the sand now turned to hard sandstone.

All these ways of preservation are interesting and enlightening when they occur, but at least 99% of all animal fossils, to be very conservative, and a great many plant fossils have the skeletons or hard parts preserved and no others. For the vertebrates what is usually preserved is the hard bone and the still harder teeth, nothing else. Sometimes cartilage is so impregnated with lime in living animals that it is as hard as bone and may be as well preserved. We used to speak of fossil bone as petrified and still do sometimes in a loose sense. We now know, however, that it is seldom petrified as our ancestors understood that word and most nonpaleontologists still do. Even in the oldest fossils the original bone substance has seldom "turned to rock," or even been replaced by some quite

---

larged thin section showing microscopic cell structure. F, footprints and rump-print of a sitting dinosaur (Triassic of the Connecticut valley) with sketch restoration of the primitive dinosaur in the position in which the prints were made (data from Lull).

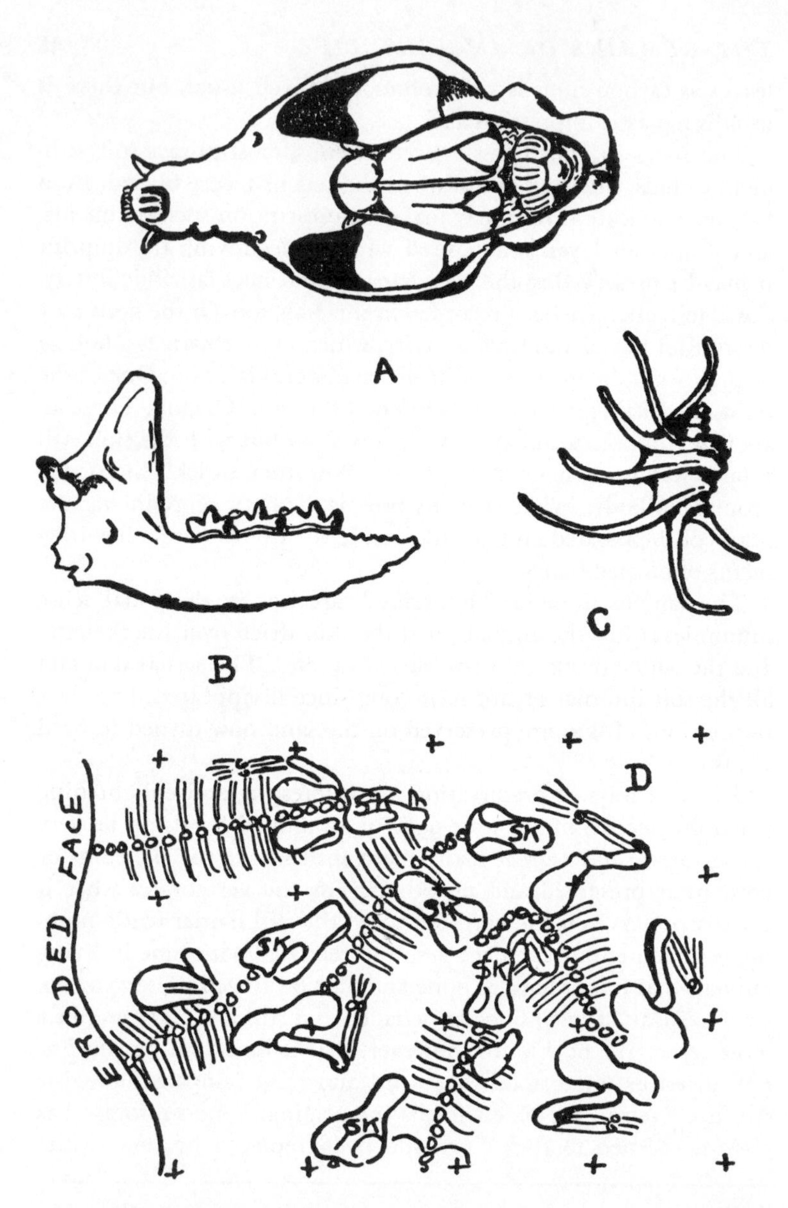

FIG. 4. Some kinds of fossils. A, skull of an early carnivore (*Cynohyaenodon,* early Oligocene) seen from above; the top of the brain case is broken away and a natural cast of its interior is exposed. B, broken lower jaw of a fossil shrew (*Domnina,* early Oligocene), more typical

different mineral. Usually the original hard material of bone and teeth which formed when the animal was alive is still there. Perhaps it is somewhat rearranged in structure but with little or no change in composition, a complex compound of lime and phosphate. The soft materials that occupy larger or smaller cavities and canals in the hard bone or tooth tissue soon decay and leave empty spaces. Sometimes that is all that occurs. More often these spaces are later filled by some mineral—silica (silicon dioxide, the substance of rock crystal or of many sands) or calcium carbonate (the substance of limestone) are the most common fillers but many others may occur. It is this filling that makes fossil bone heavier than recent bone and also more brittle. Usually the original bone or tooth substance becomes discolored. Fossil bones and teeth are rarely white, often black, and may be almost any color of the rainbow.

Most shells are made of some form of calcium carbonate, and this, too, may be preserved in fossils without change or with only microscopic recrystallization. More commonly than in the case of bones, however, this rather soluble material may be leached out, leaving a cavity in the rock, or may be replaced by some other mineral. Replacement by silica, a much harder and less soluble material than calcium carbonate, is frequent and has special value for preservation and recovery of the fossils. Some shells and the outer coatings of all crustaceans, insects, spiders, and their relatives are composed of chitin or organic compounds of similar properties, flexible but tough, something like your fingernails although of different chemical composition. Such materials are very resistant to decay and may be preserved indefinitely in fossils without marked change.

Internal cavities, such as occur in shells or in skulls, are often filled with matrix (the sediment in which the fossil is buried) or with some mineral and may be preserved after the surrounding shell or bone is dissolved or weathered away. These internal casts,

---

than A or D of usual materials for the study of fossil mammals; the original is less than half an inch long. C, a fossil sea snail (*Pterocera,* Jurassic), the shell preserved whole, as is frequent for fossil shells. D, rough quarry diagram of part of a deposit of skeletons of an early mammal (*Coryphodon,* early Eocene), marked in two-foot quadrants; six skulls (SK) and parts of associated skeletons are seen.

often called endocasts for short, are also valuable fossils which reproduce the gross forms of the soft parts that filled the cavities when the animals were alive.

The outermost tissues of plants often contain a hard waxy substance, cutin, which is resistant to decay and may preserve microscopic detail in fossils long after other tissues are carbonized or entirely lost. The woody (cellulose) tissues of plants are frequently very well preserved, even to minute features of the cells, by what is still usually called petrification. It has, however, been found, as it was for bones, that the process does not correspond with old and popular ideas of petrification. What happens first, as in the case of bones, is simply that the hollow spaces and those left by decay of soft organic materials are filled by a mineral deposit. Then the cellulose walls may carbonize or decay and their places may be filled by a slightly different mineral, but some of the organic material of the walls commonly remains even in ancient fossil plants. As with replacement and secondary filling in animal fossils, the minerals concerned are usually calcium carbonate or, especially, silica. Many other minerals may be involved, but most petrified wood is silicified.

Besides their actual remains, ancient animals have left other traces that are preserved in the rocks and are also considered fossils. The dinosaur eggs found in Mongolia are famous, and other fossil reptile and bird eggs have been discovered. Although fascinating, they are of minor scientific importance because their occurrence is so sporadic and they tell so little about the animals that laid them. The same is true of gizzard stones or gastroliths, gravel or pebbles swallowed by reptiles and birds to help grind their food. Moreover, most of the polished "dinosaur gastroliths" cherished by amateur collectors were never really inside a dinosaur or any other animal. Oddly enough, the excrement of some animals, especially carnivorous animals, fossilizes readily into a hard phosphatic mineral mass, and fossil pieces of excrement, called coprolites, are abundant. The burrows of animals of many sorts are also fairly common fossils, formed when the burrow is filled by some sediment that contrasts with their walls.

Tracks and trails of all kinds of animals, from worms to men, may also be preserved in the rocks. They are widespread and important fossils known almost throughout the long fossil record. The tracks called *Chirotherium*, which will be mentioned again later, have a long and curious history, and dinosaur tracks from many parts of the world are famous. Especially diverting are footprints found seventy years ago in hard rock in the courtyard of a prison at Carson City, Nevada, and long popularly interpreted as human. Mark Twain wrote a humorous article making them the aftermath of a drinking bout. They finally turned out to be tracks of an extinct ground sloth.

There are still other sorts of fossils, but enough has been said to suggest the usual nature and great variety of these extraordinary messages from the past.

To those who follow it, the pursuit of fossils is more exciting and rewarding than any pursuit of living fish, flesh, or fowl. It has all the elements of skill, endurance, suspense, and surprise; and the resulting trophy may be a creature never before seen by man. To be sure, fossils do not fight back—they do often seem to elude the pursuer—but I have never seen a fox riding to hounds or a lion carrying a rifle.

The question most often asked of a fossil collector, especially of the variety "bonedigger," is "How do you know where to dig?" There are so many different kinds of fossils and ways to hunt for them that a complete answer would require several books this size. The truest short answer is "You don't." One usual way to hunt for fossils follows the procedure you would have to follow if you were told to go find an object an inch long or less, exact shape and nature unknown, supposed possibly to have been dropped somewhere in an area of wilderness ten miles square. There is no esoteric sense and no instrument to tell that a fossil is buried at a given place. The collector just has to go look until he finds fragments of fossils exposed at the surface by erosion. If the fossil has entirely weathered out he picks it up. If part of it or if other fossils are still embedded there then he knows where to dig, and does so.

Although the procedure is often no more than patient search, special skill is involved. The hunter does not know beforehand where to dig, but he needs to have a good idea of likely areas in which to look for places to dig. Without knowing just what he is going to find, he has to recognize it when he does find it. Local inhabitants who are not amateurs of paleontology often walk over fossils every day of their lives without seeing them. And a fossil in the field does not look like one in a museum case. All that signals to the eye "fossil" and not "rock" may be an indefinable difference in color or the slightest oddity of shape. In a different kind of fossil hunting, fossils may even have to be collected without being seen at all. The collector has to judge what bed of rock might contain microscopic fossils and then take a sample of it back to the laboratory, perhaps a thousand miles or more away, before finding out whether he is right.

In any case a good fossil hunter must be a skilled geologist. For modern scientific purposes a fossil has no value unless its exact place in the sequence of rocks is determined. Usually, too, information as to the nature of the rock, its manner of deposition, and other geological details are required.

There is also skill in extracting fossils from their burial places without damage. There are many techniques for different circumstances, among which the most elaborate is perhaps that used in modern collecting for taking out large, brittle, and cracked skeletons. This involves careful exposure of the deposit from above, separation of the bone-bearing bed into blocks of manageable size, hardening the bones and matrix with shellac or plastics, and encasing each block in splinted plaster and bandages. The procedure is more complex than suggested by this brief summary. Experience and usually a rather elaborate set of special tools and materials are necessary for success.

Some fossils are weathered out or can be broken out in the field in such shape as to be studied without further preparation except, perhaps, soaking with a preservative. Most of them, however, require further cleaning and piecing together before all their preserved features can be well seen. In the laboratory, too, there are innumerable different techniques adapted to different kinds of fossils and the different sorts of rock in which they occur. Only a few examples can be mentioned in this brief account. Sometimes removal from the rock is best accomplished by long labor with hammer and chisel. Dental drills and other small rotary grinders or pneumatic chisels are often useful, or small fossils may be scratched out with awls and needles under a microscope.

Of great importance in recent studies of fossil shells, especially, are methods of chemical preparation. It is often possible to find parts of a limestone bed in which fossil shells are abundant and have been replaced by silica. If blocks of such rock are soaked in hydrochloric acid with appropriate precautions, the rock itself is dissolved and the fossils remain. Extraordinarily delicate features, such as hairlike spines, may be preserved by this method although lost or not even noticed if preparation is mechanical rather than chemical. Even more important for modern study of communities and populations of fossils is the fact that the method facilitates mass collecting. A large mass of rock treated chemically may produce thousands or even tens of thousands of separate fossils. Other chemical methods are also used. For instance chitinous fossils can be freed from a silica matrix with hydrofluoric acid.

In addition to cleaning fossils from the rock in which they are embedded, some groups of fossils and some sorts of studies require a large variety of additional methods. Fossil wood and some small animal fossils can be precisely identified only by means of the microscopic internal structure. This can be seen in slices so thin as to be transparent, slices that cannot be cut to this thinness in such hard material and must be ground. Sometimes an equivalent result can be obtained by polishing a surface on machinery similar to that used in gem cutting. The polished surface is then etched with acid and painted with certain plastics. When dry the plastics can be peeled off, and the peel retains an imprint of the fine structure of the fossil. Peels can also be used to follow structure all the way through a fossil such as a small skull. After each peel is pulled the surface is ground down again and the process repeated at very short intervals. From these serial peels, as they are called, enlarged wax or plaster models of the original fossil and its internal features can be made. There are also special ways of making casts of internal cavities. They can be filled with some acid-resistant material and their walls then removed by acid. Or, especially for study of brains and associated nerve patterns, a skull may be cleaned out and the inside painted with latex, which hardens to form but remains flexible enough to pulled out in one piece.

Staining of fossil tissues, radiography and ultraviolet photography, rotary milling of matrix, flotation on heavy liquids, and many other methods are also used for study and recovery of fossils of various sorts. Over the years it was learned what fossils are and how to find and collect them. Now we are learning how to extract from them all the information they can give us, which is much more extensive than our predecessors dreamed.

3. *Fossils and Rocks*

Paleontologists operate between two other sciences: geology and biology. Some paleontologists worry about whether they really are geologists or biologists and argue the point at great length. The answer is that they are more than a bit of each plus something distinct from either. Some are more interested in rocks and use fossils principally as an aid in the study of geology. Others are more interested in life, but are still geologists at least to the extent that geology is an essential aid in understanding the history of life. This book is frankly written more from the latter point of view, but the whole science of paleontology is inextricably interwoven with both geology and biology. Neither can be omitted from any study of the life of the past even though one or the other dominates different aspects. This chapter is about some of the aspects that are predominantly geological.

These aspects begin in the collecting field, where a paleontologist must operate primarily as a geologist. Fossils occur in rocks, and rocks are the business of the geologist. Not all rocks are also the paleontologist's business. A usual broad classification of types of rocks (although it is not really as clear cut as it seems) is to divide them into three major classes: igneous rocks, which have solidified from a molten state; sedimentary rocks, which were deposited on the surface of the earth's crust, not from a molten state; and metamorphic rocks, which were originally either igneous or sedimentary but have been so altered as to lose the typical features of those other two sorts of rocks. With the rarest exceptions paleontologists deal only with sedimentary rocks.

The first requirement for preservation of any fossil is that it be buried. (Exceptions will keep bobbing up, but they really have no importance; the dried-out ground sloths mentioned in the last chapter were not buried.) Continued exposure on the surface of the earth's crust, even at the bottom of the ocean, leads sooner or later to destruction by decay or by scavengers. Almost the only way for burial to occur is for the remains to be covered quickly by some sort of sediment. Igneous rocks, even in cases when they harden on the surface and not, as they often do, deep down in the crust, are usually so hot that they destroy any organic remains they may encounter. (Up bob other exceptions, but still of little importance: the rhinoceros also mentioned in the last chapter was in an igneous rock, and tree trunks are sometimes preserved as molds in lava.) The processes of change from sedimentary to metamorphic rocks destroy any fossils that may have occurred in the sediments. Because the process is gradual, poorly preserved fossils may still occur in partly metamorphosed rocks, but if metamorphism continues these are eventually destroyed.

The commonest kinds of sedimentary rocks are shales, which are hardened mud or clay, sandstones, which are hardened sand, and limestones, which may form in a variety of ways such as from lime mud, by precipitation from water, or from limy parts of animals and plants. These three kinds of rocks intergrade. There are also a great many other kinds of sedimentary rocks which likewise may intergrade with these and with each other. In fact, the sedimentary rocks are so varied and complex that their classification is one of the major and not yet wholly solved problems of geology.

Any sedimentary rocks may contain fossils, although these are much more likely to occur in some sorts than in others. For instance, fossils are generally rare in heavy conglomerates (hardened gravels with large pebbles) or in gypsum (a chemical deposit of calcium sulphate, formed under conditions in which few organisms can live). They are usually abundant in limestone, although completely unfossiliferous limestones also occur. Some sedimentary rocks are formed entirely, or nearly so, by fossils; you might almost say that they are fossils. Coal is one of these. When pure it is entirely made up of remains of fossil plants. The rock called diatomite consists of the siliceous hard parts of microscopic plants, diatoms. Many varieties of limestone are formed largely or almost entirely by hard parts of organisms. Common chalk is usually composed in great part of the limy skeletons of microscopic organisms. Coquina limestone is a mass of broken shells. Other limestones represent ancient reefs formed by corals, some types of seaweeds (calcareous algae), and other organisms. Petroleum and asphalt or other hydrocarbon deposits, which are also rocks to the geologist, are of fossil origin too. They have, however, altered chemically and migrated away from the actual remains of the organisms from which they are derived.

Close study of sedimentary rocks may reveal under what conditions they were formed and therefore whether fossils are likely to occur in them and if so what kind. This help to the fossil hunter is returned by the fact that fossils, when found, may tell even more as to the conditions under which sediments were deposited, a point to which we will return. Although limestones, for instance, form in a great many different ways and places, they are most likely to be deposited below tide level in rather shallow, warm seas away from the mouths of rivers. Some particular sorts of limestones are almost confined to these situations. Such limestones are good places to look for fossils of bottom-living shells or of corals. A hunter of, say, dinosaurs would not waste time on them. A sort of platy, thin-banded, limy shale is most likely to form in large, fresh-water lakes. Such shales are the happy hunting ground of collectors of fresh-water fossil fishes. Abundant but puzzling ancient marine fossils called graptolites are common in a type of black shale in many parts of the world and quite rare in any other sort of rock. The hunter of skeletons of land animals would expect little success (although he might make an occasional lucky find) in any of the sorts of rocks just mentioned as examples. He would look first for deposits of clays or massive shales with local lenses of sandstone.

Experience in collecting for particular sorts of fossils usually develops a feeling for the sorts of rocks most likely to contain them or a rule of association between rock types and fossils that works well in particular areas. Thus a collector of ancient mammals in western United States is hopeful when he finds badlands eroded in broadly banded clays. Moreover, it happens in some areas, at least, that some sorts of mammals are common in red clays, other sorts in gray or yellow clays. Whole skeletons are more likely to occur in ancient silts or fine sandstones and only fragments in coarse sandstones and conglomerates. Or the silts and the sandstones, even though of the same age, may contain quite different fossil mammals. Gray clays and some fine sandstones may be crowded with fossil leaves, which are extremely rare or entirely absent in red clays.

Everywhere rocks and fossils have characteristic associations or facies. This fact, unsystematically noted or only vaguely realized by earlier geologists and paleontologists, has become a fundamental principle of modern research on rocks and on fossils. As regards the fossils, especially, I will have something more to say about it in later chapters. With more particular reference to the rocks, an additional comment here may give just a hint of the interesting complications often involved in rock facies. This can be done by the example diagrammatically summarized in Fig. 5. Here there are, roughly speaking, three facies both as to rocks and as to fossils.

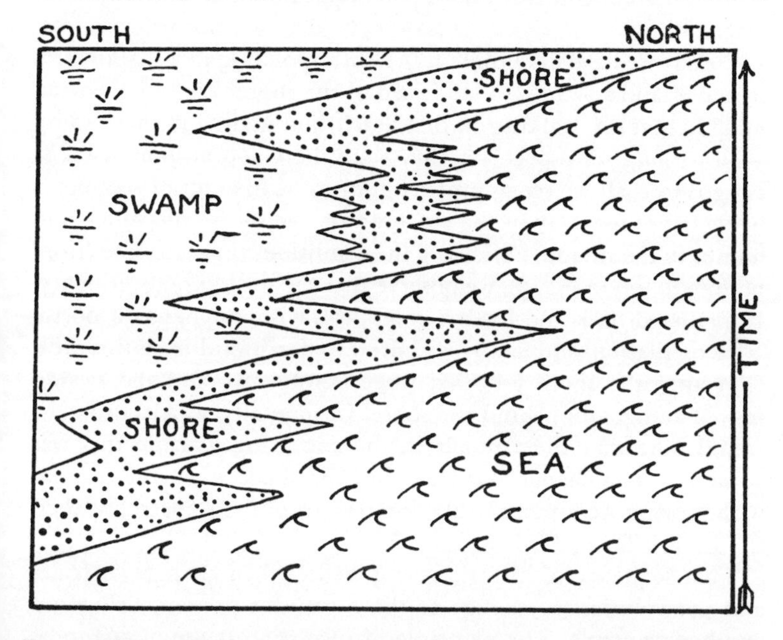

FIG. 5. Facies of sediments (and fossils) during part of late Cretaceous time along a south-north line in northwestern New Mexico. Muds were deposited in the sea, sand near the shore, and coal, mud, and sand in the swamp. These are now shales (Mancos formation), sandstones (lower sandstone members of the Mesaverde formation), and coal measures (coal members of Mesaverde), each with a different and characteristic facies of fossils. Deposition was of course from bottom to top, so that is the direction of time. Deposits on the same horizontal line were formed at the same time. (Data from W. S. Pike Jr.)

Actually, the details are considerably more complicated, as details have a way of being, but we will not let that worry us here. There is shale with abundant fossils, mostly shells, such as would occur near shore (oysters, for example) or in a shallow but open sea (such as floating ammonites, relatives of the chambered nautilus). Then there is sandstone with fewer fossils and those such as would occur along a beach: some leaves blow from inland, fragments of oyster shells, shark teeth, and other things tossed up by the waves. Finally there is a facies quite mixed up as to rocks but mostly shales and coal with abundant fossil land plants but no other fossils in most places.

Clearly we have here three facies corresponding to a shallow sea, its shore, and a swamp to landward of the shore. We can go swimming in that sea, walking on that beach, and wading in that swamp —all in long retrospect, for they were there 100 million years or so ago. In each we see appropriate forms of life and of sediments being deposited, clay in the sea, sand on the shore, and muck and humus in the swamp. Following the conditions through time (from bottom to top in Fig. 5), we observe that the sea was not stationary. It oscillated back and forth a good deal, with emphasis on northward oscillation, and finally withdrew to northward from this area. Of course the shore followed these movements, and the swamp moved along, too, behind the shore. Incidentally, a paleontologist would note that the animals in the sea changed while this was going on. The change was slow, but the shells when the sea was withdrawing were perceptibly different from their ancestors when the sea first flooded this region.

That sequence of historic events, reflected in changes of facies and in evolutionary change of the organisms present, brings up another principle. The principle of succession is the most fundamental of all as far as the present subject is concerned. We saw in the beginning of Chapter 2 how long it was before this occurred to anyone and how slow others were to accept it after Steno did think of it. That slowness can only be ascribed to blindness (and the blindness to false dogma), because the principle itself is almost ridiculously obvious and simple. It is no more than the fact that when things are laid one on top of the other the lower ones must have been there before the upper. This applies to sedimentary rocks. They are deposited on the surface of the earth's crust, and their original sequence from bottom to top has to be the order of succession in which they were laid down.

The whole science of historical geology and the whole timing of the history of life are based on that simple principle. Of course it does turn out not to be quite so simple when you come to apply it to actual cases. It is all right and infallible when applied to a truly continuous pile of flat-lying sediments in any one place: their succession along one line of traverse or section from bottom to top is unqualifiedly their sequence in time. Even in one local section, however, trouble can arise from the fact that such successions are not always truly continuous or flat-lying. Sometimes a section of the earth's crust breaks and is thrust over another. Then older rocks may be pushed over younger. At other times a segment may be tilted so that the rock layers or strata are vertical or even overturned. Then it may be difficult to be sure which way was originally up. Such occurrences are spectacular, but they need not detain us long here. They have happened only rarely and in few localities, and geologists and paleontologists have learned how to recognize them and to straighten them out. The methods of correlation, next to be discussed, usually clear up these special cases as well as more general problems. There are also a surprisingly large number of clues as to which way is up in sedimentary rocks. (A whole book of more than 500 pages was recently devoted to this peculiarly specialized aspect of geology: *Sequence in Layered Rocks,* by R. R. Shrock.) One of the most interesting is based on fossil footprints: the hollow side that took the impression of the foot must have been the surface, hence the down or ground side, when the sediments were soft.

There are more serious and basic problems of sequence and contemporaneity in study of the history of the earth and its life. Most of these arise from the facts that the rock sequence is not complete in any place and that it differs greatly from place to place. No single pile of sedimentary rocks represents the whole of geological time. The rock sequence as we can actually study it represents at any one place only such times as sediments were being deposited there. Usually this is only a limited span of geological time. There are also interruptions in the record. Within a sequence sedimentation may stop for an appreciable interval. Some sediments already deposited may also be removed by erosion before new sediments are added to the pile. In order to be available for extensive study the rocks deep in the pile also have to be uplifted and exposed at the surface again by erosion, which may entirely remove some of the top layers of the pile. Deep well drilling may tell much about unexposed lower parts of a pile of sedimentary rocks, but wells only penetrate a few thousand feet and do not permit really thorough knowledge of buried strata.

The whole sequence, then, has to be pieced together from parts observed a bit here and another bit there. In this piecing two more serious difficulties arise. The same kind of sedimentary rock or even what appears to be exactly the same bed of rock may be of quite different ages in different places. On the other hand, sediments deposited at the same time may be quite different in character. On a small scale these effects are both visible in the example given above (Fig. 5). Marine shales, shore sands, and swamp shales and coal were all being deposited at the same time in different places. If the shore sands are traced across country they can be followed continuously, but they are not of the same age throughout. In general they are decidedly younger in the north than in the south.

Similar variations, with almost endless differences in details and combinations, occur wherever sedimentary rocks are followed far across country and wherever the sequences of different regions are compared. As a result it is quite impossible to build up the whole sequence of earth history or even a really complete sequence for any long subdivision of it on the basis of the rocks alone. This is another problem that our predecessors were slow to grasp. It was involved in a series of great controversies among geologists in the latter years of the 18th and earlier part of the 19th centuries. One school of thought, the "Neptunists," was led by Abraham Gottlob Werner of Freiberg who taught that all rocks had been formed one after the other in a definite order, first granites, then slates and shales, then limestone and sandstone—even Werner did realize that the sequence is more complicated than that, but that was his general idea. Moreover, Werner thought that at any one time the same sort of rock was deposited all over the earth, which thus was everywhere covered with the same successive layers, like an onion.

You would think that geologists had only to look around them to see that Werner's theory could not possibly be true. Sediments are being deposited today, but obviously they are being deposited only in certain places and regions and not in others. Moreover, they obviously are different in different places, here mud, there sand, and another place lime. The same stream can even be seen to deposit fine mud and coarse sand only a few feet or inches apart.

The reason why some early geologists failed to apply these obvious facts to interpretation of the rocks was simply that they did not bother to observe the facts. Others, however, thought that such observation would have no bearing, or would be downright sinful. They believed that the sedimentary rocks were the record of a great catastrophe (the Biblical deluge, of course) or of a whole series of catastrophes unlike anything to be seen today. They found it emotionally impossible or thought it sacrilegious to look squarely at facts that contradicted faith and tradition.

Others thought that the present is the key to the past and that the same sorts of forces have acted throughout geological history. If we see a river eroding its banks it seems logical that the same sort of erosion over longer periods of time probably carved out the river valley. If all kinds of sediments are now deposited at the same time in different places, this must always have been possible. It was acceptance of this view, the principle of uniformitarianism, that finally made a true science of geology. That acceptance came gradually but can conveniently be dated from the publication in 1830–33 of Sir Charles Lyell's three volumes grandly entitled *Principles of Geology: Being an Attempt to Explain the Former Changes of the Earth's Surface, by References to Causes now in Operation.*

At the same time that geologists were beginning to realize that they could not build up a geological time sequence from the rocks alone, they were finding out how this can be done: by the use of fossils. At each particular age in later earth history there were widespread, characteristic animals and plants, and different times had different animals and plants. These facts still are the basis of the greatest contribution of paleontology to geology. They lead to solutions for both the basic problems mentioned above. Rocks in different regions and perhaps of different sorts are of the same age if they contain the same species of fossils. The correct sequence in time can be pieced together by sequences of fossils.

In New Mexico, where we took a walk a couple of chapters ago, we saw varicolored banded clays and sandstones in which were remains of animals called *Coryphodon* and *Hyracotherium.* In England in the London Clay, very different in appearance and thousands of miles away, fossils of the same two animals have been found. In neither place are there exactly similar fossils from rocks distinctly higher or lower in the local series. The London Clay of England and the San José formation of New Mexico are thus demonstrated to be of the same age, an age designated as early Eocene. This sort of demonstration is known as "correlation" in geology, and it is a large part of the work of the paleontologist in aid of geology.

Piecing together of the sequential record extends the usefulness of correlation. Suppose in one place is found a series of strata containing in succession from bottom to top fossils (or associated groups of fossils, faunas or floras) *a, b, c, d.* (See Fig. 6.) In another

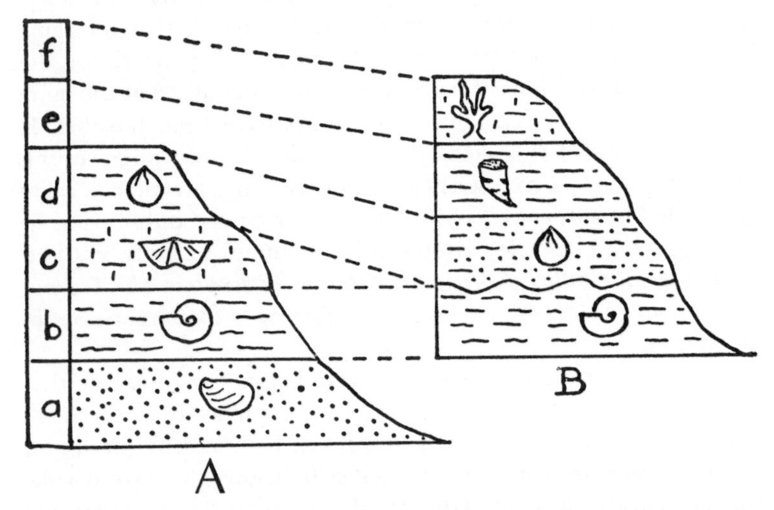

FIG. 6. Correlation and sequence. A and B are sequences of fossil-bearing rocks in two different places. Correlation is shown by characteristic fossils. (In practice the successive fossils would be less distinctive, and whole faunas, not single fossils, would be used.) The whole time and faunal sequence, a–f, is shown to the left. The wavy line in B is an unconformity.

place there is a sequence with fossils *b, d, e, f.* The rocks with *b* and *d* can be correlated in the two places. It is then clear that rocks and fossils fit into a more extensive whole time sequence represented by *a, b, c, d, e, f.* Another interesting and important fact will be noticed. At the second locality *c* is missing. That means that at the particular time when fossil (or associated fossils) *c* was living sediments were not being deposited there, or if they were deposited they were soon removed by erosion. In geological terms there is here an unconformity under the bed containing *d* and a span of time not here represented by rocks.

The problems of correlation and sequence are now also studied by a number of other methods. Over short stretches of space and time the character of the rocks is helpful, and recently some very ingenious physical aids have been devised, such as an instrument for measuring the radioactivity of different rocks pierced by an oil well. Nevertheless, the general solution of these problems has always depended mainly on fossils, still does, and probably always will. Of course the use of fossils in this way is usually much more complicated than has been suggested in this summary. A major complication is that fossils, like rocks, are not everywhere the same at any one time. They also have facies. A marine sediment will have no fossils in common with, say, valley sands deposited at the same time. Fossils also have geographic differences, as animals and plants do today. For instance, the plants and land animals of the geological epoch called the Miocene are markedly different in North and in South America. Broad scale correlation is still rather inexact in many cases, but increased knowledge and specific attention to these difficulties is providing better and better solutions for them.

The building up of a time scale for earth history has required that names be applied to subdivisions of the scale. By use of such names geologists and paleontologists can most conveniently designate where in the sequence their rocks and fossils belong. The subdivision and nomenclature now generally adopted started with Werner, William Smith, and especially Lyell. They have been greatly modified and refined by further study since Lyell. The present system subdivides the whole of geological time into eras, the eras into periods, the periods into epochs, and the epochs into ages, a descending scale like this:

ERA
- PERIOD
  - EPOCH
    - AGE

Each subdivision has a name. Because of the difficulty of exact long-distance correlation, the smallest subdivisions, the ages, still have different names in different regions. The same age names are not, for instance, generally applied to North American as to European rocks and fossils. But in some unusually favorable cases the European names can now be accurately applied in America, and it is probable that a uniform world-wide nomenclature can eventually be achieved. Except for later epochs, those of the Cenozoic era, different epoch names are also now usually applied on different continents. Essentially the same period and era names are now used by geologists everywhere. Paleontologists, especially, usually do not bother with subdivisions of early geologic time before fossils were abundant and varied enough to make them widely useful in correlation. For our purposes, then, those early eras (although three or more times as long as all geologic time since then) may be lumped together as "pre-Cambrian."

The names that are in general use are given in Table 1. In spite of some pedagogical methods to the contrary, I think that an excellent command of paleontology requires little rote learning. This table is an exception, however. Anyone who is really interested in rocks or in fossils, in the history of the earth or of life, should learn these names and their sequence by heart. I recommend learning them from the bottom upward: that is the natural sequence from older to younger or, in sedimentary rocks, from lower to higher.

You will note that approximate ages in millions of years are given for the various subdivisions. You may wonder why geologists and paleontologists bother with the sequential system of names and do not simply give the ages of rocks and fossils in years. The reason is that in the present state of knowledge even the best established year dates in geological time are exceedingly rough approximations. If I say that a certain fossil (such as *Hyracotherium*) is early Eocene in age, everyone knows just where it fits in the sequence. Also anyone else studying the same fossil would also label it "early Eocene." But if I said its age was 60 million years no one would be sure whether it belonged before or after the middle Paleocene fossils, for instance, and some other students might say with equal assurance that it was 65, 50, or 40 million years old.

A great many ways of estimating geological ages in years have been proposed and tried. Biblical exegesis led Bishop Usher to believe that the earth was created in 4004 B.C., a date still accepted by the most naïve of the faithful although nothing in the Bible

TABLE 1. THE GEOLOGIC TIME SCALE

MILLIONS OF YEARS SINCE BEGINNING (VERY ROUGH ESTIMATES)	ERAS	PERIODS	EPOCHS
.02	Cenozoic *	Quaternary	Recent
1			Pleistocene
10		Tertiary	Pliocene *
30			Miocene *
40			Oligocene
60			Eocene
75			Paleocene †
135	Mesozoic	Cretaceous	(Epoch names not world-wide in application)
165		Jurassic	
205		Triassic	
230	Paleozoic	Permian	
280		Carboniferous ‡	
325		Devonian	
360		Silurian	
425		Ordovician	
500		Cambrian	
More than 2,000	Pre-Cambrian	(Period subdivision not well established)	

* A few European students display their knowledge of the original Greek roots by spelling these names "Kainozoic" or "Cainozoic," "Pleiocene," "Meiocene." This is an affectation of which the ancient Greeks would not have approved.

† In Europe, where the Paleocene is poorly represented, some geologists still lump it with the Eocene. The Paleocene was as long as the average for Tertiary epochs and surely deserves recognition, which is now given it by all American and most European students.

‡ American geologists usually call the early Carboniferous "Mississippian" and the later Carboniferous "Pennsylvanian," but these names have not come into general use elsewhere.

really so much as suggests it. More rational methods, such as one based on how rapidly the sea is becoming saltier, gave more reasonable dates, running well up into the millions of years but still certainly too small. Modern estimates are based on radioactivity in certain minerals. This radioactivity results in breakdown of old and production of new elements at very constant and known rates.

The details have nothing directly to do with fossils and I do not propose to spend space on them here, but the interested reader will find them in some of the books listed at the end of this one. The method is still very inaccurate because of the scarcity of fully suitable minerals and the great difficulty of entirely adequate analysis. We may, however, be sure that we now have the right orders of magnitude. We do not know certainly whether the Cenozoic era began 50 or 80 million years ago, but we are quite sure that the right figure is somewhere around those and not, say, around 3 million as some geologists thought as recently as 1920 or so.

Before leaving the subject of fossils in their relationships to rocks something should be said about the use of fossils in economic geology. Such uses are many and essential. The practical study of almost all sedimentary or stratified rocks is likely to involve fossils and may depend largely and directly on them. Fossils are therefore generally useful in seeking and exploiting all economic resources associated with such rocks. Such resources are very extensive and some of them constitute the very basis of industrial civilization. Petroleum and coal are the two most evident and probably now, at least, most important of these. In some areas, however, and increasingly so in others the most precious and indispensable resource is water, and the search for water often is aided by fossils. Many other important resources also come from stratified rocks. It may suffice to mention that such rocks are at present the principal source of uranium and other radioactive elements in the United States. In these deposits the occurrence of ore is intimately related to fossil distribution.

The most extensive directly economic use of fossils is now in the petroleum industry, which employs the great majority of all paleontologists not engaged in teaching or basic research. Here paleontologists play the general role of aid in mapping and interpreting sedimentary rocks, and they have also a particularly specialized role. In the rational development of an oil field it is absolutely essential to know exactly what rock strata are pierced at each level in drilling a well. There are many different clues useful for this difficult task. Several different methods are commonly employed at the same time, but one of the best and most widely used is based on fossils. Ordinary drilling methods break the rock into small fragments before it is brought to the surface, destroying almost all fossils large enough to study with the naked eye. Even drilling that brings up a core of solid rock—drilling so costly that it is a sort of last resort in most commercial operations—seldom produces enough large fossils to be extensively useful. There has therefore grown up, mainly from the needs of the petroleum industry, a subscience of micropaleontology, rather oddly distinguished from other sorts of paleontology by the fact that the fossils it studies are smaller and can more often be recovered from oil wells.

These microfossils are of many sorts: numerous kinds of one-celled plants and animals or, better, protistans (see the appendix); included are foraminiferans, radiolarians, infusorians, flagellates, diatoms; tiny, shelled crustaceans (ostracods); and parts of larger animals such as spicules of sponges and some other animals, worm jaws (scolecodonts), fish scales, and odd tooth- or jawlike structures, apparently not really teeth or jaws and of unknown origin (conodonts). Of these microfossils, by far the most generally useful are the creatures called Foraminifera or, affectionately by their students, "forams." Most micropaleontologists are foram specialists. In many oil fields forams are abundant in drill cuttings and different forams or, with greater precision, different assemblages of foram species are characteristic of different strata. By this means, within a given oil field correlation of beds can be made from well to well. Distribution of petroleum in the field often depends on underground geologic structure, on how the deeply buried strata are tilted or folded. By noting the depths at which beds with characteristic forams occur in various wells, this structure can be accurately determined even when no trace of it is visible on the surface. (Fig. 7.)

Aside from such special applications of the general techniques of correlation and succession, study of associations of fossils may also have commercial bearings. For instance, much petroleum is now being recovered from ancient reefs. Extensive studies of the organic, community structures of these and of similar recent reefs are now being made, with confidence that this will assist the search for petroleum.

One other, as yet commercially unimportant application of fossils to the study of rocks will be mentioned, just as an example of how diverse and sometimes unexpected such applications may be. Rocks in the earth's crust are often twisted, squeezed, or

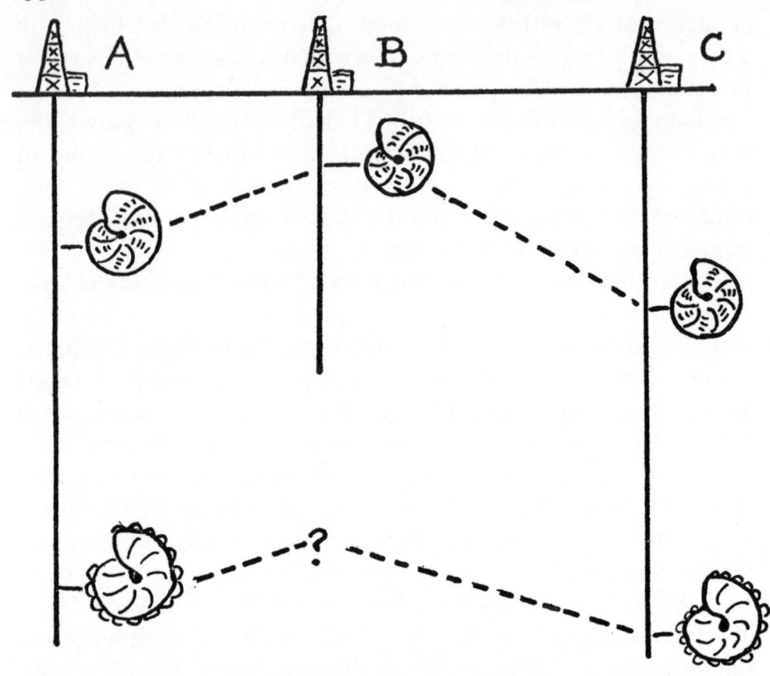

FIG. 7. Subsurface correlation by Foraminifera from oil wells. A, B, and C are three wells in the same field, which has two subsurface zones indicated by characteristic forams, shown greatly enlarged. The fact that the upper zone occurs at less depth in B than in A or C indicates an upward flexure of the rocks. In general this would be a favorable indication for finding oil in B, which has not yet been drilled to the lower zone.

sheared in various directions. The amount and direction of such deformation have an important bearing on many geological problems, including economic applications. These details are very difficult to determine because the shape of the rock before deformation is not known. But the shape of a fossil before deformation may be precisely determinable: for instance, the right and left sides of many animals are symmetrical. If such fossils occur in deformed rocks the exact amount and direction of deformation can be determined. (Fig. 8.)

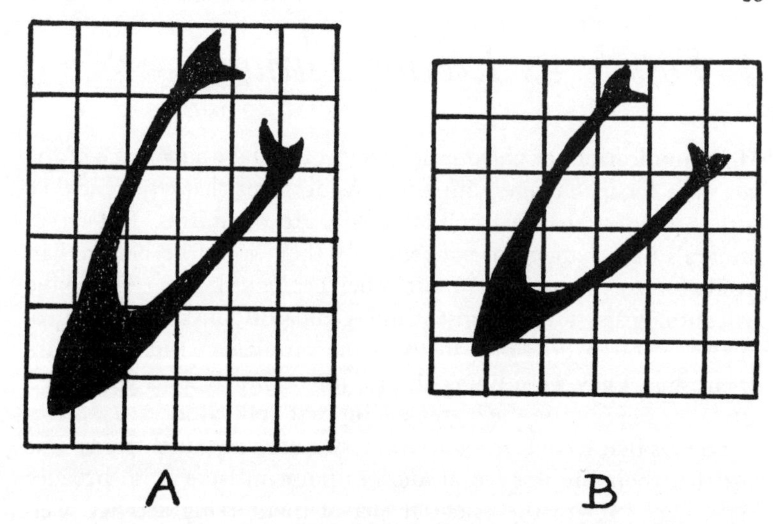

FIG. 8. Measurement of deformation by means of a fossil. A, the lower jaw of a fossil bird (*Protornis,* Oligocene) as found in the rock. It is distorted but must have been symmetrical when buried. B, same jaw with its symmetry restored. Comparison of the grids in A and B shows how much the rock has been deformed and in what directions. (After F. Stüssi.)

4. *Fossils as Living Things*

MOST applications of paleontology to geology take advantage of the fact that fossils change with time. Aside from that, they involve little attention to the fact that fossils are organisms. Indeed, if minerals had changed so conveniently they would be better than fossils for correlation and most other geological uses. (Economic paleontologists, mineralogists, and geological physicists, or geophysicists, often do work on the same problems.) But fossils *are* organisms. They were living things, just as much so as you and I are now. Therein lie their greatest interest and value.

To eyes that have learned to see, fossils are very much alive. The fact that their life was led at another point in time is, in broader view, irrelevant to the essential fact of their living activity. Certainly they are as much alive as any frog or cat on the dissecting table of a biology class. Animals that died yesterday and those that died millions of years ago are equally dead, and were equally alive. Unfortunately, however, those that died millions of years ago are not so completely preserved. This makes for greater difficulty in interpretation of their structure, in seeing it in relationship to living function, and calls for great skill and insight. (There is an advantage appreciated by students of animals longer dead: they have no smell.)

It was noted in Chapter 2 how seldom a fossil represents a whole animal. Whole fossil plants are equally rare. Leaves, seeds, wood, and roots, if preserved at all, are frequently preserved separately. Associating them correctly is one of the special puzzles for paleobotanists. Animal fossils usually consist of hard parts only. In the case of fossil shells, corals, and the like, all the hard parts of an individual can usually be found. This is commonly true of fossil fishes, also. Their bony skeletons are composed of many separate parts, loosely or not tied together once the soft tissues have decayed, but they are frequently buried whole. Then their whole skeletons are preserved and may be beautifully displayed on a slab of rock.

In dealing with fossil land animals, especially the higher vertebrate forms, reptiles, birds, and mammals, the paleontologist is faced with an added difficulty. These animals usually die in the open air and are buried, if at all, after scavengers have torn them apart or the soft tissues have all decayed and the loose bones have been scattered by the elements. If and when they are buried this is likely to be a result of a flood or freshet, which further breaks and scatters the remains. Then the fossil collector rarely finds them until they are weathering out of the rocks, when even highly mineralized bones soon disintegrate into meaningless fragments or are washed away.

The result of all these hazards is that complete skeletons of the extinct higher vertebrates are relatively very rare. *Hyracotherium* or eohippus is a common fossil. Literally tens of thousands of single teeth, jaw and skull fragments, and separate bones of the skeleton have been found. Since 1838, when the first fragments of *Hyracotherium* were found, every museum has coveted a skeleton of this famous ancestor of the horse, and hundreds of collectors have searched diligently. But today (at the beginning of 1952) there are only four skeletons in museums, and none of these is really complete.

Occasionally a lucky accident will preserve whole skeletons or even many of these together. A dust storm, a sudden flood, a mud hole, a caving stream bank, or the like may kill and bury animals forthwith, precluding scattering of skeletons. In the work of my own museum department we have recently found one deposit with dozens of complete skeletons of a small Triassic dinosaur (*Coelophysis*), previously known from a few tiny scraps, and another with fourteen skeletons, five of them absolutely complete, of an Eocene mammal (*Coryphodon*) of which only one poorly preserved skeleton had been known. Such incidents stand out in a fossil hunter's life because they happen so seldom. Ninety-nine per cent and more of the fossil dinosaurs or mammals found are represented only by fragments. In exceedingly few of the extinct species of these groups are all of the hard parts known.

Although he had more casual predecessors, the first man to work extensively and really systematically on the paleontology of fossil vertebrates was Baron Georges Léopold Chrétien Fréderic Dagobert Cuvier (1769–1832), "Cuvier" for short. He found that almost any single part of the skeleton of an animal is characteristic. It is unusual for any two species, even though closely related to each other, to have a single bone or tooth *precisely* alike in the two. Consequently it usually suffices to find one bone or one tooth to determine which species you have. This extremely fortunate circumstance makes it possible to follow the history of mammals, for instance, in great detail even though most of the fossils concerned are mere fragments.

Cuvier's observation had, however, a less happy result, one that follows and annoys paleontologists to this day. Because single bones and teeth are usually characteristic and because skeletons are integral units with the bones and teeth all coordinated and working together, Cuvier thought that a single piece should suffice to tell what the whole skeleton is like. Unfortunately this is not true, or at least if it is true paleontologists have not yet learned the trick. Cuvier was barely in his grave when one of the most spectacular refutations of "Cuvier's law" of correlation of anatomical parts turned up. He had concluded that certain sorts of teeth always are correlated with occurrence of hoofs on the feet of the same animals. But a whole group of extinct animals (that of the chalicotheres) turned out to have great claws on the feet and yet to have teeth such as *must* (by law!) be associated with hoofs. This baffled the paleontologists of the time and enraged some of them. It has sobered paleontologists ever since.

I mentioned before that old errors never die. This one never has. Ever since Cuvier it has been common knowledge that a paleontologist can reconstruct whole skeletons from a single bone. Today many people know this who know nothing else about paleontology. Alas! everyone knows it but the paleontologists. They can often (not always) *identify* a fossil animal from a single bone; they cannot *reconstruct* it unless most of the skeleton is objectively known. A single horse tooth is very characteristic and easy to identify offhand for anyone who has ever really looked at one. (Oddly enough, few horse fanciers seem ever to have done so; they are always bringing horse teeth to the museum to be identified.) Skeletons of horses are well known, of course, and if you see a horse tooth you know it belonged to an animal with such a skeleton. Even so, you could not really reconstruct its own individual skeleton and get, for instance, the length and slenderness of the cannon bones right.

Similarly, you might pick up a fossil tooth from the Eocene. By comparing it with all more or less similar teeth previously found from that epoch, you might find that it was very like one of them and could then identify it as belonging to the same species. It might, for instance, be like one of the five teeth in a little jaw fragment that my predecessor, the late Walter Granger, found in Wyoming and named *Shoshonius cooperi*. You could further fit it into its place in history and establish, for instance, that its nearest living relative is *Tarsius spectrum*, the quaint little spectral tarsier, a sort of pre-monkey, in southeastern Asia. But you could not reconstruct its skeleton, or even its skull. These parts were surely not closely like those of living *Tarsius*. The teeth are not exactly alike, and considerable change must have occurred in the intervening 50 million years or so. The skull and skeleton are unknown in *Shoshonius cooperi* or any closely related Eocene animal, and they simply cannot be reconstructed on any properly scientific basis now established.

The reconstruction of a whole fossil animal requires, then, for a start that all or nearly all the hard parts be found. We have seen that this is usual for some fossils, such as oysters, or fishes, and quite unusual for others, such as mammals. Even for the latter, however, the fossil hunters have been at it long and diligently enough to have accumulated a fair series of skeletons as samples of the more important different sorts. Even these are very rarely absolutely complete, but reconstruction of some missing parts is possible without undue guesswork. Since most vertebrates are nearly symmetrical, a bone missing on one side, only, can be supplied by a plaster replica, in mirror image, modeled after the other side. Two or more partial skeletons of the same species may be at hand. Parts missing on one and present in another can safely be cast or modeled from the latter. Some missing parts can be inferred with sufficient exactness from adjacent parts. A missing section of a rib will merely bridge between the pieces that are present, and a missing whole rib must (with, as usual, occasional exceptions) be much like preceding and following ribs.

The next step in bringing a fossil back to life is to consider what the soft parts were like. In almost all cases these are gone forever. But they were fitted around or within the hard parts. Many of them also were attached to the hard parts, and usually such attachments are visible as depressed or elevated areas, ridges or grooves, smooth or rough patches on the hard parts. The muscles most important for the activities of the animal and most evident in the appearance of the living animal are those attached to the hard parts and possible to reconstruct from their attachments. (Fig. 9.) Much can be learned about a vanished brain from the inside of the skull in which it was lodged.

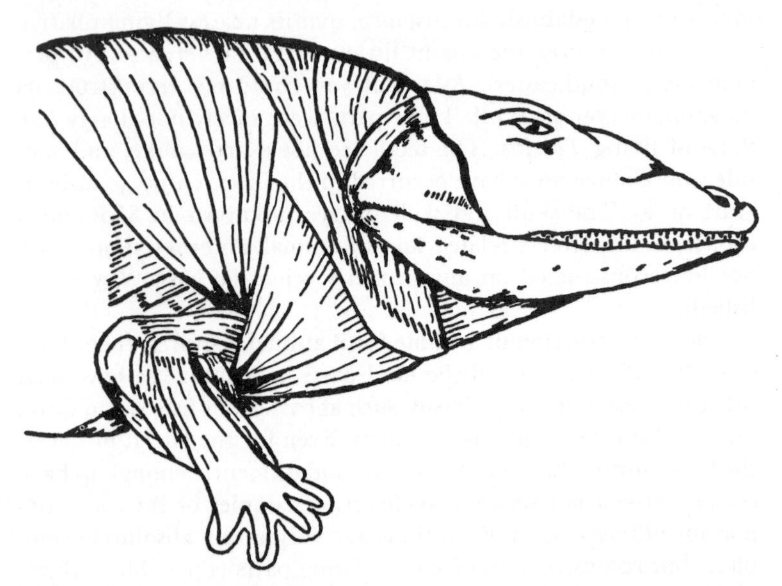

FIG. 9. Restoration of some shoulder and leg muscles of a fossil amphibian (*Eryops,* Permian). (After R. W. Miner.)

Restoration of the external appearance of an extinct animal has little or no scientific value. It does not even help in inferring what the activities of the living animal were, how fast it could run, what its food was, or such other conclusions as are important for the history of life. However, what most people want to know about extinct animals is what they looked like when they were alive. Paleontologists, who are people too, also would like to know, and they try to oblige. Things like fossil shells present no great problem as a rule, because the hard parts are external when the animal is alive and the outer appearance is actually preserved in the fossils. The color is usually guesswork, although color bands and patterns are occasionally preserved even in very ancient fossil shells.

Animals in which the skeleton is internal present great problems of restoration, and honest restorers admit that they often embody considerable guessing. The general shape and contours of the body are fixed by the skeleton and by muscles attached to the skeleton, but surface features, which may give the animal its really characteristic look, are seldom restorable with any real probability of accuracy. When possible, the present helps to interpret the past. An extinct animal presumably looked more or less like its living relatives, if it has any. This, however, may be quite equivocal. Extinct members of the horse family are usually restored to look somewhat like the most familiar living horses, domestic horses and their closest wild relatives. It is, however, possible and even probable that many extinct horses were striped like zebras. Others probably had patterns not present in any living members of the family. If lions and tigers were extinct they would be restored so as to look exactly alike. No living elephants have much hair, and mammoths, which are extinct elephants, would doubtless be restored as hairless if we did not know that they had thick, woolly coats. We know this only because mammoths are so recently extinct that prehistoric men drew pictures of them and that the hide and hair have actually been found in a few specimens. For older extinct animals we have no such clues.

Length of hair, length and shape of ears, color and color pattern, presence or absence of a camel-like hump, and many other features are shaky inferences if not downright guesses in most restorations of fossils, especially those of mammals. Other restorations of ancient animals, such as those of ichthyosaurs or of many fishes, are certainly very near their original appearance, although even in these color and pattern are almost always unknown. At worst, a restoration by a good artist with competent scientific collaboration shows what the animals *may* have looked like, and may be enjoyed without taking it too seriously.

In fossil plants the whole external form is commonly preserved. The only serious problem is, then, finding and associating all the parts. When this has been done careful restoration may be completely accurate except, again, for details of color and color pattern.

Without attempting a restoration much may be learned about the life activities of ancient animals from their hard parts, from shells and other external supports or from reconstructed internal skeletons. In fact even single teeth or parts of dentitions and parts of skeletons too incomplete for reconstruction may permit some valid and useful inferences about the living animals. The problems and the possible inferences are so extremely complex and varied, there are so many different sorts of fossils each with a different way of life, that only a few examples can be given here.

One of the most abundant groups of fossils and one of the most important both for geological correlation and for study of principles of evolution is that of the ammonites. These were extinct marine mollusks, distantly related to the living squids and octopuses and more closely related to the chambered nautilus. Like the latter they had an external shell divided by partitions into separate chambers. As the animal grew it added new chambers and lived in the last of these. Older chambers were usually, at least, filled with buoyant gases. Most ammonite shells were simply coiled, more or less like a nautilus shell. Others were straight, or partly coiled and partly straight, or spiral, or they had various strange and complex forms.

Paleontologists used to take this variety of shapes for granted or else to consider it as exemplifying a mysterious (and, I feel sure, quite fictitious) "racial senility." More recently some students have made a good beginning toward understanding the differences of shape as reflections of different ways of life. Most ammonites, those with the usual type of coiling, could rise or sink in sea water at will and feed or rest at different levels. Among the others, some were heavy and must have lived on the bottom, others so light they must always have stayed near the surface. Separation and partial straightening of the coils made the shells more stable and kept them near a fixed orientation, often with the aperture pointed upward or upward and forward. Some spirally coiled and straight shells must have floated with the aperture directed downward or downward and forward. (Fig. 10.) These factors must have been related, at least in a general way, to different food habits.

Many fossil pelecypods (or lamellibranchs), the clams and their innumerable bivalved relatives, are so like recent relatives in general form and structure that they must have had similar habits, but there are some striking exceptions. Among these are the extinct forms collectively called *Gryphaea.* Instead of having the two shells nearly alike, as is usual in clams and, to less extent, in the oysters from which *Gryphaea* was derived, these animals had one shell decidedly bigger and heavier and growing in a tight coil,

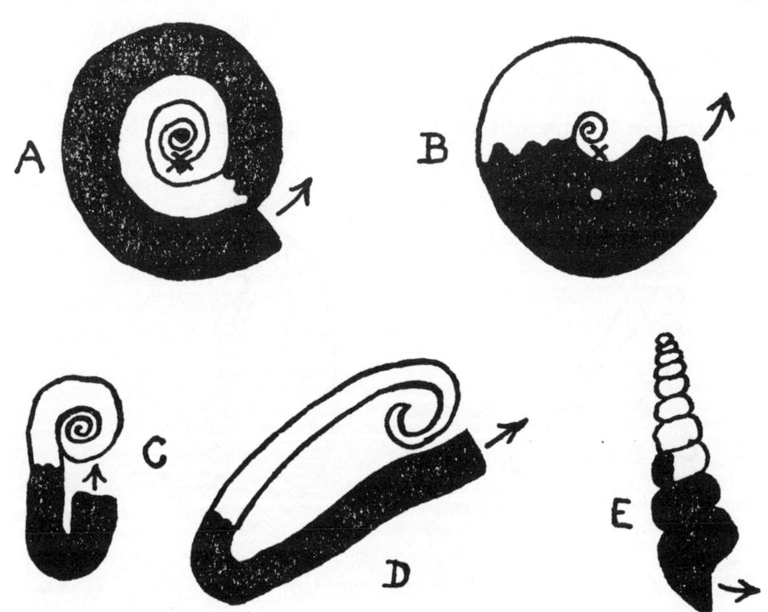

FIG. 10. Diagrams of the floating position of some ammonites. In A and B a cross indicates the center of buoyancy and a dot the center of gravity. The shell floated with the center of buoyancy above that of gravity and was more stable the farther the distance between the two. A, unstable coiled form that could float in almost any position (*Dactylioceras*). B, stable coiled form (*Sigaloceras*). C–E, highly stable forms. C, with aperture pointing upward (*Macroscaphites*). D, with aperture pointing forward and upward (*Lytocrioceras*). E, with aperture at bottom, pointing forward (*Turrilites*). In all, the chamber occupied by the living animal is black. The white parts, which were divided by partitions not shown, were filled with buoyant gases. Arrows indicate the direction of the opening. (After Trueman.)

while the other remained relatively small, light, and flat. This oddity used also to be ascribed to some mysterious (and probably quite nonexistent) evolutionary force. Now it is reasonably interpreted as evidence of the animals' special habits. They thrived particularly on the surface of loose or muddy bottoms, a difficult environment for most pelecypods. The odd shell pattern kept the aperture pointed upward while the animals were attached and oriented it in a livable position when they broke loose and lay on the bottom. (Fig. 11.)

Fishes are animals adapted to moving about, often with great rapidity, in a fluid but resistant medium. Their multiplicity of

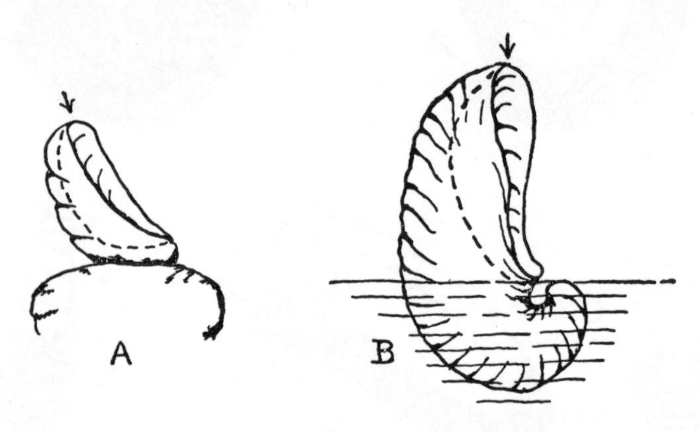

FIG. 11. Adaptive significance of coiling in an extinct oyster (*Gryphaea*, Jurassic). A, an early stage of coiling, animal attached by one shell to a hard object. B, an advanced stage, animal broken free and sunk in mud, held in living position by its shape and weight. The arrows indicate the point of greatest opening of the shells, which must be kept free for survival. The broken lines indicate the approximate shape and size of the living chamber, to the right of the line.

body and fin forms lends itself particularly well to mechanical analysis in both fossil and recent fishes. Fishes flattened from back to belly usually feed or rest on the bottom in shallow water, and those with spindle-like, streamlined bodies are fast, free swimmers. Swimming mechanisms vary from wavelike lateral undulation of the whole body to sculling or fanning with flexible fins attached to a completely rigid body. Hydrostatic adjustment of weight relative to that of water, of centers of gravity and of buoyancy, and of placing and mechanisms of fins are extremely varied and may be very delicately adjusted. Certain tadpole-like early fishes, pteraspids, were heavier than water and lacked fins that could serve as elevating planes. They could, nevertheless, swim free of the bottom because the shape of the head was such as to produce an upward thrust if the animal moved forward with the head higher than the tail. This was associated with the fact that the stiffening axis of the tail fin was turned downward instead of being medial or upturned as in all recent fishes. (Fig. 12.) (I use "fishes" here in a broad,

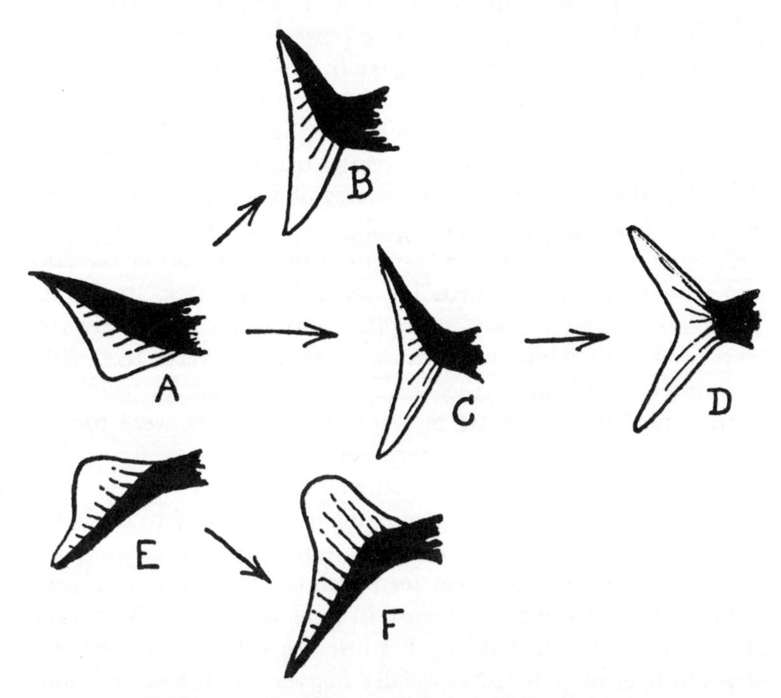

FIG. 12. Evolution of the tail fin of fishes. The relatively stiff power axis is in black; the fin proper is a more flexible sculling blade. Sequence A–B is associated with presence of anterior elevating fins and relatively low buoyancy; A–C–D is associated with elevators and increase of buoyancy; E–F is associated with absence of elevating fins. A, placoderms, some sharks. B, specialized "shark" or chondrichthyan. C, early bony fishes. D, later and most recent bony fishes. E, pteraspids, some anaspids. F, anaspid.

popular sense; really there are four quite distinct groups of "fishes" and the pteraspids were Agnatha. See the appendix.)

Food habits of extinct mammals can be judged in a general way and sometimes very specifically from their teeth. Most fossil mammals with well-developed canine teeth and shearing posterior teeth ate meat by preference. If they had sharp, large canines, only moderately heavy or light jaws and posterior teeth, and were swiftly running or leaping forms, they were predaceous, that is, they captured living prey (like cats or dogs today). If the teeth were heavier and blunter, the jaws more powerful, and the limbs less agile, they probably ate carrion (like hyenas). Mammals with low-crowned teeth and more or less numerous, nonshearing tooth points or cusps (like ourselves or like pigs) generally were omnivorous, which does not mean literally that they could eat everything but that they ate a great variety of nutritious animal and vegetable food, small animals, grubs, seeds, fruit, etc. Mammals with some sort of cropping apparatus at the front end of the jaws and with heavy, ridged grinding teeth farther back ate plants. Those with relatively low teeth (like our cows) ate mostly soft leaves and twigs, while those with high, cement-covered teeth (like our horses) could eat harsh grasses. Land mammals in which the teeth tended to degenerate or were lost altogether were for the most part those eating ants or termites (like living anteaters). And so on, through a long series of detailed inferences as to food habits.

In such cases the control of deduction is the study of recent mammals with mechanically similar teeth. Inference is of course more difficult and less secure when fossil animals had structures quite unlike any known in recent forms. In spite of this difficulty reasonable inference may be possible. For instance, certain of the extinct, elephant-like mastodonts, especially *Platybelodon,* had very long lower jaws, at the front end of which were two tusks flattened like the leading edge of a shovel, followed by a depression in the jawbone like the scoop of a shovel. Nothing quite like this occurs in any recent animal. Careful study by H. F. Osborn and by W. K. Gregory of the specimens themselves and of the anatomy and habits of recent forms even remotely similar has worked out the functional relationships beyond much doubt. The shovel-tusks probably spaded up mud and soft dirt in which were succulent tubers and other vegetable food. A short, heavy trunk helped to root for these tidbits. The jaw-scoop lodged a specialized, mobile anterior part of the tongue which selected food items and moved them to the posterior part of the mouth. There a heavier, less flexible part of the tongue pushed the food between the stout posterior teeth, where it was crushed, and then pushed it down into the throat and helped to swallow it. (Fig. 13.)

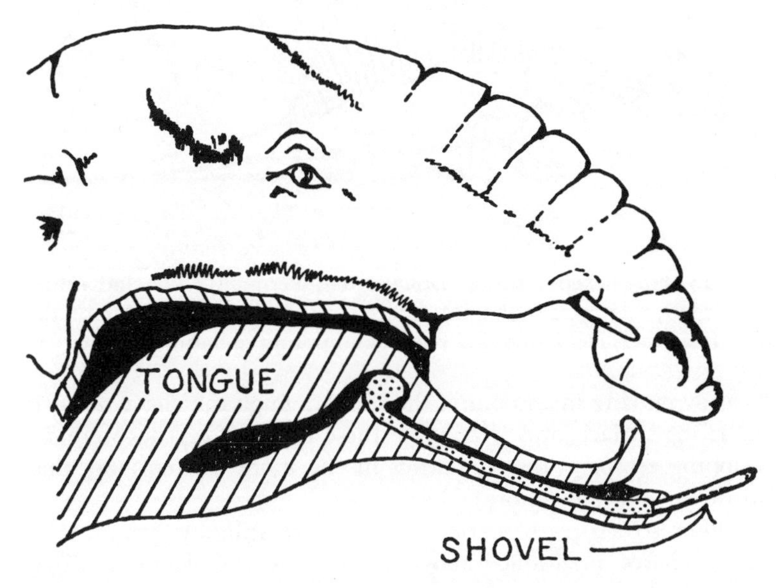

FIG. 13. Head of a shovel-tusked mastodont (*Platybelodon*, late Miocene of Mongolia). The upper part of the head is shown in external view, restored. The lower part is shown as if sliced down the middle. The stippled part is bone, a slice through the scooplike bone joining the two sides of the jaw anteriorly. (After Gregory, but much modified.)

An even greater enigma that may now be solved is provided by some of the Permian reptiles with high spines on the vertebrae, evidently supporting a tremendous dorsal fin or sail. (Fig. 14.) Many conjectures were made about this structure, varying from the guess that it was simply a fantastic nuisance (which is not entirely explanatory!) to the even less likely guess that the animal was aquatic and used the sail for tacking upstream in a brisk breeze. Now detailed study, mostly by A. S. Romer, strongly suggests that the "fin" was a radiator and heat receptor. By appropriate exposure to the sun and regulation of the large blood supply, body heat could be either gained or lost and thus maintained at a relatively constant effective level. Modern reptiles do regulate their body

FIG. 14. Fin-backed reptiles (*Dimetrodon,* Permian). The individual to the left, sidewards to the sun, is receiving maximum radiation on the fin. The other, facing the sun, receives minimum radiation.

heat by moving in and out of the sun, although they have no such radiating surface, and relatives of these Permian reptiles were developing temperature regulation in the more efficient way that made them our ancestors.

Other structures of extinct animals are still mysterious. The chalicotheres, mentioned above as violators of Cuvier's law, had very peculiar feet terminating in large claws. (Fig. 15). The mechanical operation of the feet could be and has been worked out in detail, but how the animal really used them remains a puzzle. The most popular conjectures are that the claws were used to dig up soft, buried tubers or alternatively that they were used to hook the animal to a tree trunk while it ate leaves. There are serious objections to both views, and no good way to choose between them. In honest fact the use of the feet is unknown. Yet they must have been effectively useful, for the group lasted a long time, some 40 million years, and spread over most of the world. Moreover, another group in South America, to which chalicotheres did not spread, independently evolved rather similar feet. But no living animals have anything of the sort.

The most satisfactory sort of study of the whole life activity of an extinct animal can be illustrated by one of the last dinosaurs, *Triceratops.* Numerous skulls and several nearly complete skeletons have been found in western United States, especially in Wyoming and Montana. The skeleton was already fairly well reconstructed in 1891 by O. C. Marsh, although corrections in details

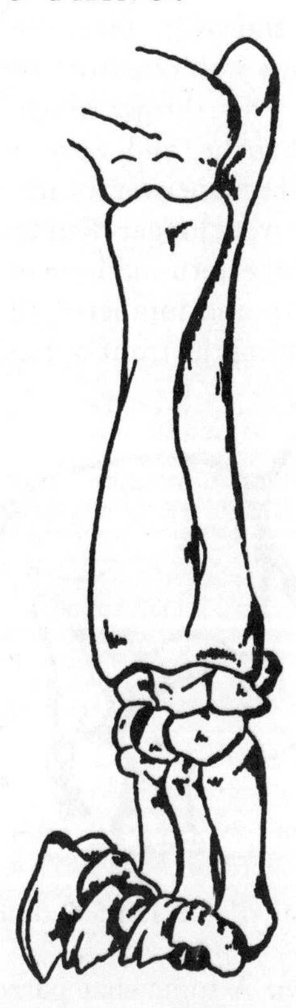

FIG. 15. Bones of the forearm and foot of a clawed ungulate (a chalicothere, *Ancylotherium,* early Pliocene). (After Schaub.)

of anatomy, proportions, and pose have since been made. The animal is large (about 25 feet long in large adults), heavy, four-footed, with a relatively short but stout tail, and with a tremendous head up to 8 feet long with a bony frill extending back over the neck, two large, forward-pointing horns over the eyes, and a smaller horn on the nose. (Fig. 16.) The most important things to know about a living animal are manner of locomotion, food and general metabolism, sensory and nervous equipment, reproduction, means of defense and offense, and habitat. All these facts can be reasonably inferred for *Triceratops*. Let us briefly review them:

Locomotion: The hind legs are elephantine and pillar-like, upper segments longer than lower segments. The front legs were not pillar-like (although so shown in some restorations), but were carried akimbo, with elbows well out from the body and feet turned back under the body. This throws much weight on upper arm muscles, which needed to be, and were, exceptionally powerful, but also makes for greater maneuverability of the front of the body and the head. Hind legs were longer than forelegs and the feet were elephantine, with hooflike terminations of the toes. The normal gait must have been slow and lumbering but with possibility of a quick lunge and of pivoting the front of the body rapidly.

FIG. 16. Restoration of the horned dinosaur *Triceratops.*

Food and metabolism: A somewhat parrot-like beak served for cropping and a very large, scissors-like tooth battery in the middle and posterior parts of the jaw for chopping. The whole apparatus suggests a diet of land plants, probably coarse and abrasive to judge by the severe wear on the dental battery, which was constantly renewed throughout life. Enormous quantities of food must have been needed. The animal was almost certainly cold-blooded, and maintenance of livable body heat in so large a bulk in the open air must have depended on warm, little-variable external temperatures.

Sensory and nervous equipment: The eyes were large and the ears normally reptilian; the sense of smell present but relative importance dubious. The brain was ridiculously small for so large an animal, but probably no more so than would be expected from brain-body ratios in recent reptiles. It was typically reptilian in structure, and its reactions, as in living reptiles, must have been more instinctive than what we are pleased to call intelligent. An enlargement of the spinal canal between the hips suggests that the maneuvering of legs and tail was more reflex than fully consciously controlled.

Reproduction: As a land reptile, *Triceratops* would be expected to lay eggs as do most recent land reptiles. Its relative *Protoceratops* is known to have laid eggs in clutches in holes scooped out in warm sand, and *Triceratops* almost certainly did the same.

Defense and offense: The hide was surely tough, but its exact nature is unknown. There may have been bony plates embedded in it, but this remains doubtful. Defense was mainly by pivoting to present the head, which carries horns and tremendously powerful bony protection extending well back over the neck. The great skull was balanced and moved by very powerful muscles from neck and trunk to the underside of the frill, which have been restored in detail. (Fig. 17.) The head was normally carried bent down, with the brow-horns pointing forward. It is shown as held too high in many restorations; the normal pose can be determined from the organs of equilibration, semicircular canals, in the inner ear. The horns were weapons of offense as well as defense and were apparently used in combat, possibly sexual, with other members of the same species.

Habitat: From the animal itself it would be inferred to have lived on dry land in a warm, even climate and to frequent open country or glades with abundant low, tough vegetation. Its associations with other animals and plants and with types of sediments agree with this conclusion. Its ecological role in the late Cretaceous was that of the dominant, large, dry-land herbivore.

Even in such brief summary, that is much information about the life of an animal dead some 75 million years, and as much or more can be learned about many other extinct animals.

Besides their actual remains, animals have left many traces of their activities in the rocks. Some of these were mentioned in Chapter 2. Many of them are rather rare and scattered occurrences. For this or other reasons they may not make significant contributions to the history of life, even though they have a fascination as curiosities. For instance, the fairly common occurrence of fossil bones that were bitten or gnawed while fresh seldom signifies much more than the commonplace that carnivores used to bite and rodents to gnaw bones in the past as they do today. Yet the occur-

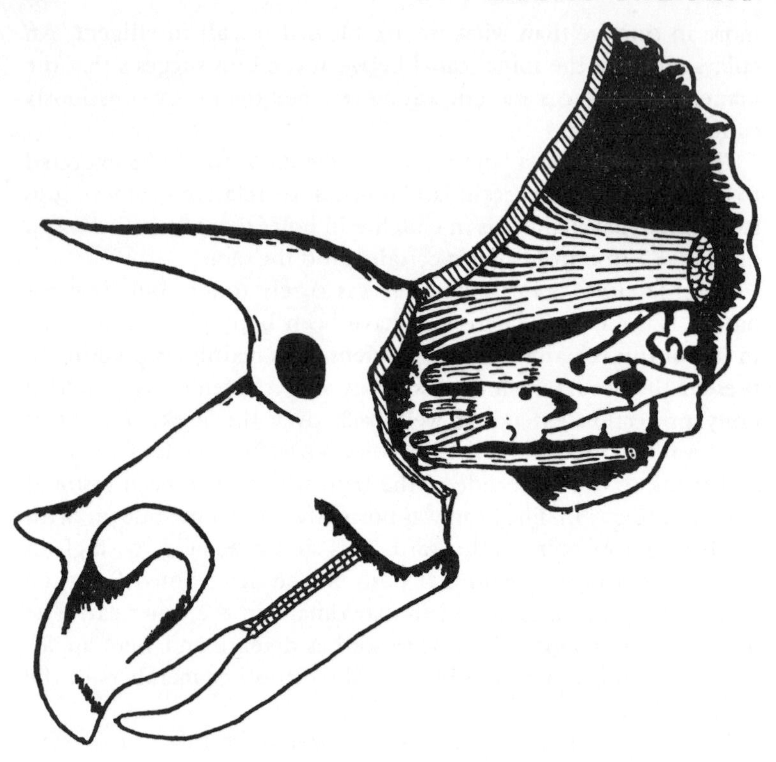

FIG. 17. Skull and part of its musculature, *Triceratops.* The left side of the frill has been cut away, revealing the anterior neck bones, and some of the muscles that support and turn the massive head are shown. The muscles are diagrammatic and in life were probably bulkier, filling the whole space under the frill. (After Lull.)

rence of a gnawed bone in beds (early Paleocene) older than any known rodents did importantly confirm the inference that an older group (Multituberculata) had filled the place of rodents in the economy of nature before the rodents evolved. And few can help feeling that the past has greater reality when they see a bone bitten by a carnivorous dinosaur 135 or so million years ago.

An ancient beaver burrow containing the skeleton of a small carnivore tells a dramatic story of an ousted or in all likelihood eaten rodent, and of retribution in the form of a freshet that trapped the carnivore in its pre-empted home. This also illustrates the difficulty and controversial nature of some interpretations of the past. These beaver burrows, as most students now believe them to be, are spiral structures, often six feet or more in length, with a straight projection at the bottom. (Fig. 18.) As preserved, they are filled with matrix harder and of different consistency than that

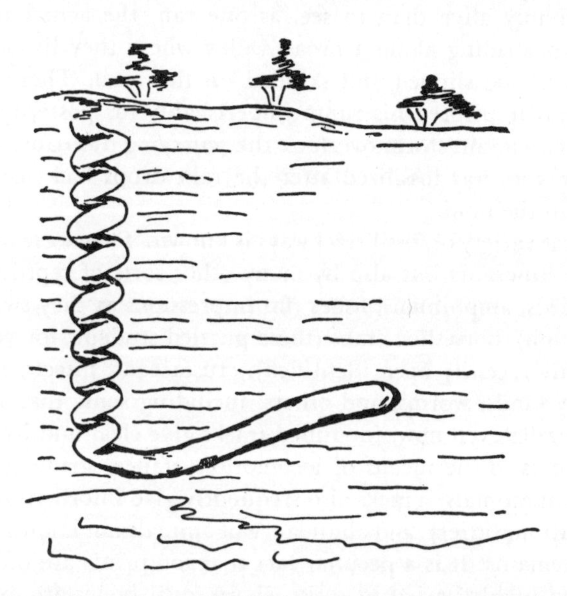

Fig. 18. A *Daemonelix* or Devil's corkscrew. This object, which is probably the burrow of a small extinct beaver, is shown as exposed in the wall of an excavation.

around them. They sometimes contain a quantity of rather formless plant material, as recent rodent burrows also do. Some students have thought that they were, indeed, fossil plants rather than burrows, and controversy about them raged for many years. Remains of a small ancestral beaver, *Steneofiber,* have been found in them, and the old view that these beavers made them now predominates. They are best known from the Miocene of Nebraska, where the settlers' name for them, "devil's corkscrews," has been learnedly latinized to *Daemonelix* by the professors at the state university. Similar, and similarly disputed, structures have also been found in the Oligocene of Europe and elsewhere.

Coprolites (fossil excrement, see Chapter 2) are abundant but so far have not been very informative, perhaps only because no one has been inspired to study them in an intensive and systematic way.

The most informative and to most observers the most impressive of traces of ancient activities are footprints and trackways. They are the actual records not of dead things but of life being lived. They might be called "fossilized actions." Nothing makes the past more vividly alive than to see, as one can, the actual tracks of dinosaurs striding along a broad valley where they lived so long ago. Here one slipped and skidded on the mud. There one sat down and imprinted his rump imperishably for posterity. In another place a rain storm overtook the scurrying dinosaurs, and the weather, too, was fossilized after the rain drops had made small craters in the mud.

A great variety of fossil trackways is known. They were made not only by dinosaurs but also by many other sorts of reptiles, mammals, birds, amphibians, fishes (fin impressions as they swam near the bottom), horseshoe crabs (these puzzled students for years and have only recently been identified), crustaceans, insects, mollusks of many kinds, worms, and others, including some that are complete puzzles even now. Identified tracks give clear and irrefutable evidence as to the means of locomotion, stance, stride, gaits, etc. of ancient animals. Tracks also frequently give information on the distribution, variety, and abundance of animals not known by their actual remains. It is a peculiar fact that footprints are often common and highly varied in rocks where fossil bones are extremely rare, notably in Triassic rocks in Europe, North America, and elsewhere.

Some of these Triassic prints have been particularly famous and objects of much speculation and research. Tracks called *Chirotherium* were first noticed by scientists in Europe in 1834 and have since been found widespread over the earth. (Fig. 19A.) Some of them look surprisingly like human hands, although we now know that it is the left hind foot that looks like a right hand. Early conjectures assigned them to apes or baboons, but in 1841 Sir Richard Owen convinced his colleagues that they were made by early amphibians. In 1855 Lyell produced a restoration of an animal making the tracks. (Fig. 19B.) It was not then known that early amphibians were not at all like frogs in body build, so Lyell sketched a gigantic frog. He could not believe that a left foot could look like a right hand, so he had the frog cross his feet at every step as he crept forward.

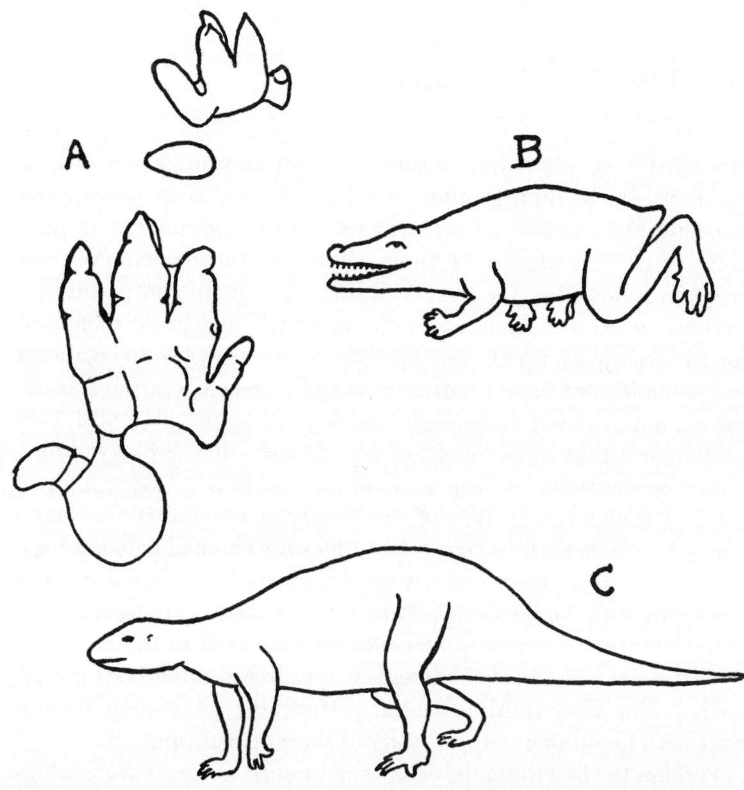

FIG. 19. *Chirotherium*. A, left fore and hind footprints as found (after Peabody). B, Lyell's conception of the animal that made the footprints, a sort of large frog walking cross-legged. C, a modern conception of the animal that made the footprints (essentially after Soergel).

Paleontologists soon learned that Lyell's *Chirotherium* is an impossible monstrosity, but the mystery of the prints was not solved until 1925. Then a German paleontologist, W. Soergel, worked out methods of inference on which really critical and scientific study of footprints, or rather of whole trackways, is possible. Such study might now claim to be a branch of paleontology in its own right. *Chirotherium* turns out not to be an amphibian after all, but a reptile, an earlier relative of the dinosaurs. (Fig. 19C.) Bones of the animal have not yet been found, but the group to which it belonged, Pseudosuchia, is soundly established.

5. *Ancient Communities*

No animal or plant lives alone or is self-sustaining. All live in communities including other members of their own species and also a number, usually a very large variety, of other sorts of animals and plants. The quest to be alone is indeed a futile one, never successfully followed in the history of life. All animals and plants also depend on external sources for the materials that build them and for the energy by which they operate. Students of life have always had some idea of these relationships and in recent years are examining them with increasing care and ingenious new methods. This particular aspect of the study of life is now called ecology, which might be translated as "the study of togetherness." The ecology of the past is an essential part of the history of life and is also being studied with increasing care, often under the name of paleoecology. Of course paleoecology and ecology are parts of the same thing and not really different sciences. They are separated only because the materials are differently preserved and presented in the two cases so that some differences in practical training and methods are required. The same thing is true of paleontology as a whole in contrast with neontology or the study of recent organisms.

Perhaps the first thing needed for an understanding of the ecology of life in the past is some idea of the physical environment in which forms now fossils lived. Everyone is aware of the fact that physical environments differ greatly and that the differences are reflected by the organisms living in each. The broadest separation may be indicated as that between marine, fresh-water, and land environments. There are intergradations between these. The tide zone is transitional between sea and land, a brackish estuary between sea and fresh water, a swamp between fresh water and land. Each is also extremely varied and includes many subdivisions. In fresh-water environments lakes and streams have quite different living conditions, as do small and large or warm and cold lakes, or rapid and sluggish or clear and muddy streams. On land, climate makes for the most obvious differences. Land climates are infinitely varied and depend on a large number of factors. Annual average, extremes, and distribution through the year are involved for such varied things as radiation from the sun, air temperature, precipitation, humidity, evaporation, and air movement. Then there are such environmental factors as the character of the land, level or mountainous, with deep soil, sandy, or rocky, and so on. Elevation makes for great environmental differences in land life because of associated climatic and other factors. In all regions with great contrasts in elevation both plant and animal life shows a zoning by altitude, which may be surprisingly clear cut. (Fig. 20A.)

The sea, too, has fairly well-defined depth zones. (Fig. 20B.) Usually distinguished are the littoral zone (between tides or a bit above and below them), the neritic zone (to a depth of about 600 feet, which is about the depth of effective wave action and extreme penetration of light), the bathyal zone (from about 600 to about 6000 feet), and the abyssal zone, below about 6000 feet. In all zones there is an important environmental difference according to whether organisms live on the bottom (such organisms make up the benthos), float passively and usually near the surface (plankton), or swim freely at variable depths (nekton).

In the case of ancient life few factors of the physical environment can be directly observed. Marine fossils are collected on dry land. The sea in which they lived has long departed, and its depth, temperature, motion, or salinity cannot be measured. Much can be learned from rocks about the environment in which they were formed, a point discussed and exemplified in Chapter 3. Even more can be learned from the fossils, and of course a complete and competent study takes in both rocks and fossils as well as their relationships to each other.

An example of such combined study is provided by occurrence of ammonites, the extinct relatives of the chambered nautilus mentioned in the last chapter, in the Cretaceous of Texas and Mexico. Both rocks and fossils can be rather clearly zoned, and the zones correspond approximately to depths in the ancient sea. (Fig. 21.) Four zones have been recognized within the neritic and bathyal regions. (Two more zones, in the littoral region, occur but do not commonly contain ammonites; an abyssal zone was not present in these relatively shallow seas.) Each zone has characteristic sorts and associations of ammonites. This association, once established, can be used to judge the probable depths of other deposits in which similar ammonites occur.

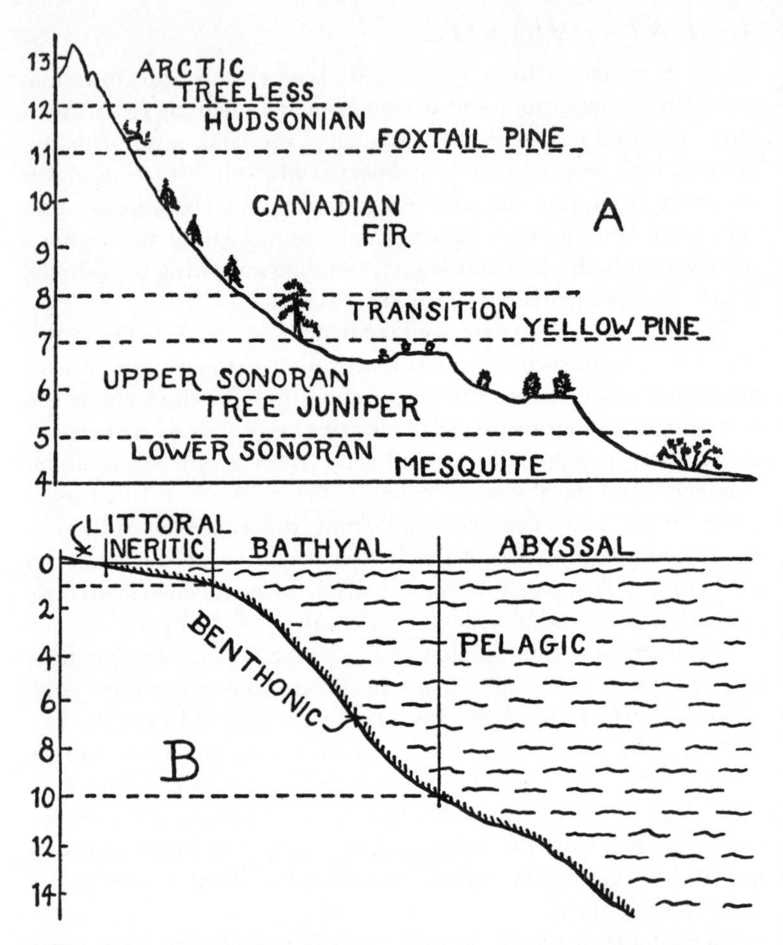

FIG. 20. Life zones. A, land life zones. The scale to the left is in thousands of feet above sea level. Land zones differ greatly from place to place; the zoning here shown occurs in New Mexico and depends on altitude, temperature (decreasing with altitude) and precipitation (increasing with altitude). Many plants and animals are so zoned; only one typical tree or shrub is here shown in each zone. B, marine life zones. The scale to the left is in hundreds of fathoms below mean sea level. Marine zones are relatively little varied and the system here shown is applied everywhere. Occupants of the pelagic zone include plankton and nekton. In a past situation like A few or no fossils would be found from above the Sonoran. Few marine fossils are known from the deeper bathyal or the abyssal zones.

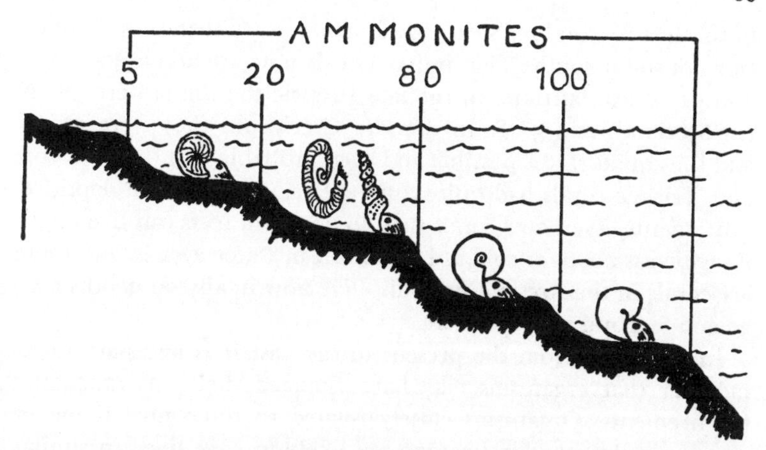

FIG. 21. Zoning of ammonites in the Cretaceous sea of Texas. The figures are depths in fathoms; depth is not scaled. (Data from Gayle Scott.)

Even without identification of individual fossils the general aspect of a fauna or flora may be indicative of environmental conditions. In dealing with fossil mollusks their variety and the average size of individuals may be enlightening. Decrease in variety and in size suggests decreasing salinity of water. Decrease in variety and increase in size accompany colder water. Both variety and size tend to be highest somewhere near the top of the neritic zone and to decrease from there toward the shore and toward deeper water. Similarly, in dealing with fossil plants the average size of leaves, of whatever precise species, may be an indication of climate, with larger leaves usually indicating moister and small leaves drier climates.

Recent corals may be good indicators of depth and temperature of water. Reef corals will not grow in water deeper than about 300 feet, and actual building of reefs occurs only at depths less than about 150 feet. Most reef corals can survive temperatures from about 65° to 95° Fahrenheit, but active reef building usually occurs between 75° and 85°. It is usually assumed that fossil reef corals required similar conditions. This is probably correct for reefs from the Jurassic onward, which were composed of corals more or less like those of today. An interesting point is that through much of this history coral reefs extended farther north than they do today but not so far south. The limits of reefs now are about 35° north and 32° south latitude. In the late Jurassic the limits were 58° N (in Scotland) and 5° S. In the late Cretaceous they were 50° N and the equator. It is possible and even probable that the Paleozoic and Triassic corals had quite different temperature and depth requirements. Even now corals that do not form reefs can live at 35° F., which is about as cold as bottom ocean water ever is, and there are corals in the Arctic. Corals also live now in abyssal depths near 20,000 feet and perhaps more.

In reasoning from the present to the past it is necessary to remember that organisms may have changed their environmental requirements. Organisms closely related to those now living or closely similar in adaptive type are likely to have lived in similar environments, but exceptions must be expected. For instance, it sometimes happens that animals now living in cool climates or water have fossil relatives that lived in what other evidence strongly suggests were considerably warmer circumstances. This illustrates what seems to be a rather general rule, although it has many exceptions: plants and animals are more rigidly confined by lower than by upper limits of temperature, water supply, etc. Thus corals that also tolerate low water temperatures may be found growing on and near reefs, but reef corals do not spread to colder water. Plants of arid or semiarid regions may thrive when well watered, but plants from damp situations cannot live under semiarid conditions.

By taking such relationships into account and by studying the *whole* fauna or flora, past environmental changes can sometimes be followed in much detail. This has been done, for instance, for the Cenozoic marine faunas of the Pacific coast of North America. (Fig. 22.) The result shows that at any given latitude the water has tended to become rather steadily colder since the Eocene. (Paleocene faunas are too poorly known there to be included.) Or, put in other words, the water temperature zones have tended to move southward. This agrees with much other evidence that the tropical and subtropical zones, although not then warmer, formerly extended much farther north than they do now. In general plants are better climatic indicators than animals. Cenozoic floral zones throughout the northern hemisphere also show greater extension

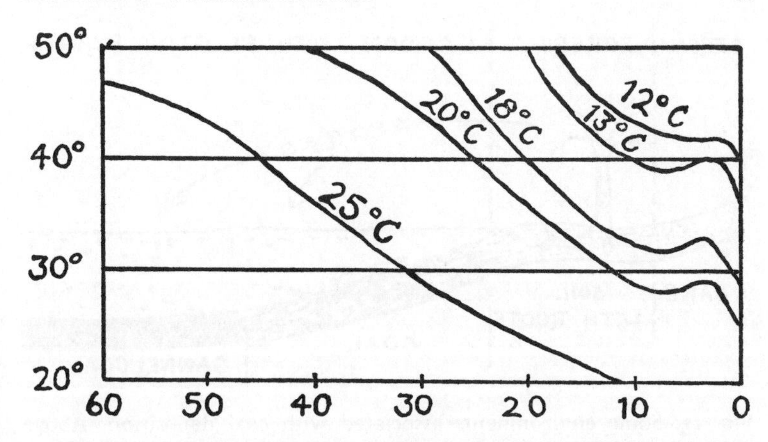

FIG. 22. Temperature of sea water on the Pacific coast of North America during the past 60 million years. The scale at the bottom is millions of years before the present; 60 million years is taken as approximately the beginning of the Eocene. The scale at the left is degrees of north latitude. The curves show water temperature in degrees centigrade. (After Durham.)

northward in the Eocene and rather steady movement southward during that era.

On a smaller scale the distribution and changes of local environmental conditions can often be followed in detail. Fossil reefs, for instance, may clearly show the lagoonal deposit behind the live reef, the growth zone of the reef proper, and its steep outer side merging seaward into the conditions of a normal neritic zone. Coal deposits may similarly show gradation from land detritus through the zone of growth of the coal forest and then a series of changing deposits into deeper water and farther from dry land in a swamp or lagoon. (Fig. 23.) In both examples each local environment has characteristic fossils not only as regards the sorts of organisms present but also in respect to their manner of preservation and burial. For instance, in the reef lagoon may be found organisms favoring calm water and high concentration of salts as well as much-broken fragments of reef animals, limy mud, and chemical deposits. The reef deposit includes corals or other reef builders preserved just as they grew, and the seaward slope has broken and detached blocks from the reef.

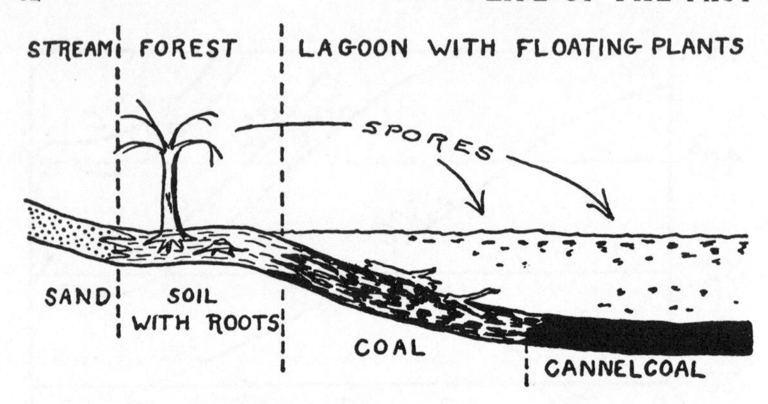

FIG. 23. Some environments associated with coal deposition. Above, environments. Below, type of deposit being formed. (Suggested by a diagram by Duparque, but much modified.)

That last remark brings up the point that the surroundings of a fossil do not necessarily represent the place and conditions under which the animal or plant lived. A fossil has gone through an experience not involved in its life activities: it has died and been buried. Death may be the result of an abnormal fluctuation of the environment. This is especially likely to be true of "mass mortality," resulting in a rich fossil deposit. Unusual cold, "bloom" of poisonous micro-organisms, and increase of salinity of water are known to cause mass mortality at present and inferred to have done so often in the past. The conditions surrounding the animals when they died were then certainly not their normal environment but quite the contrary, conditions under which they could not live. Similarly, the masses of sabertooth and other remains in the California tar (really asphalt) pits have not led anyone to believe that asphalt was a usual or needed feature of those creatures' environments.

After death, remains are often transported and finally buried in quite a different environment. Floating (planktonic) or swimming (nektonic) aquatic organisms cannot be buried where they live. They can become fossils only if their remains sink to the bottom and are buried together with those of organisms (benthonic) that really lived there. In marine deposits bottom-living shells, floating forams, swimming fishes and reptiles, and both swimming and flying birds may occur together as fossils, as they do, for instance, in the Niobrara Chalk of Kansas. Dinosaurs have several times been found in rocks certainly deposited in open seas. It has even been claimed that these dinosaurs were marine animals. Their whole functional anatomy speaks conclusively against this and so does the fact that the same and closely related dinosaurs are much more common in deposits laid down on land. It is clear that burial at sea resulted only on the rare occasions when their bloated bodies floated down to the sea in a river or were washed from a beach.

The ecologists, who happen to be especially fond of inventing fancy Greek names for the things they talk about, distinguish the thanatocenose and the biocenose. A thanatocenose is an association of dead organisms and a biocenose of live ones. Fossils found together are always a thanatocenose. They may represent a mixed sampling of several biocenoses, as obviously does the Niobrara fauna mentioned in the last paragraph. In this case, and usually in others, it requires only a little common sense, knowledge of functional anatomy, and acquaintance with recent faunal associations to sort out the several biocenoses. In other cases fossil associations represent only one biocenose or a few that are closely associated and interlocking. A good example of that kind will be analyzed briefly at the end of this chapter, after a little more background has been sketched in.

A fossil association never is an adequate sample of a whole biocenose. Virtually all biocenoses include bacteria, which everywhere play an important, indeed absolutely essential role in ecology, but bacteria are extremely rare as fossils. Many other very small or soft-bodied forms are also rare or generally absent in the fossil record although important in living communities: the various sorts of worms and many insects, for example. Bats and birds are rare as fossils in most deposits even when and where they must have been important in the biocenose. Plants and animals are seldom both adequately sampled in a single fossil deposit, although both always occur in any normal biocenose. Study of biocenoses of the past thus has to take into account a considerable number of organisms that you know were there, even though they are not there now as fossils. Or analysis may include only part of a biocenose but still may be significant and enlightening.

In every biocenose each sort of organism has its own role which differs, if only slightly, from the role of any other sort. The basic unit is a local population of each species, a group all with essentially the same ecological role and interbreeding with each other so that the group is maintained continuously over a longer or shorter time. The whole community is made up of a number, sometimes a very large number, of such specific populations all occupying the same general area and environment and all interacting on each other in various ways, directly or indirectly.

Two or more populations with the same or even in some respects essentially similar roles cannot long live together in the same community. This is the principle of ecological incompatibility. It applies generally when two sorts of organisms require the same thing (such as light, food, water, or even living space), the supply of which could be completely utilized by one of them. For instance, if two species of carnivores both prey exclusively on the same herbivores one or the other species of carnivore must change its food habits, leave the country, or die. That is the reason why every specific population in a biocenose has a somewhat different role, or in other words has a different adaptation, from any other.

These roles may often be clearly related to one of the most fundamental aspects of a biocenose: the flow of food and energy. With exceptions of no real importance all energy used by life originates in the sun and comes to the earth as radiation. It is captured and turned into chemical energy by plants, a fact made evident, although only in a superficial sort of way, if a plant is burned: the light and heat released are solar energy reconverted to radiation. The material needs of life are met by substances available in the atmosphere, the surface waters, and the thin outermost parts of the earth's crust: carbon, oxygen, hydrogen, nitrogen, and a considerable number of other chemical elements equally essential but required in much smaller amounts. From plants the energy passes on through a longer or shorter chain of other organisms. It is dissipated as it goes along and finally ceases to be available for further use by life, which nevertheless keeps going because new energy is constantly arriving from the sun. Some of the chemical materials for life are lost or made unavailable (for instance, in coal or limestone deeply buried in the earth's crust), and some new materials are added (for instance, in volcanic gases from deep sources).

By and large, however, these materials continue in an endless cycle through the different forms of life.

Plants form the beginning (as far as life is concerned) of the energy flow and an important step in the material cycle—the latter because they bring lifeless materials such as carbonic acid gas and water back into organic forms usable by other organisms. A next step is taken by the very numerous animals that eat plants. Even in one community the variety of plant-eaters may be large because each sort specializes on certain plants or parts of them. Next steps are those involving animals that eat other animals. There may be several or many steps here because animal-eating animals are themselves eaten by other animals, and so on. In each step from the plants on, "eating" means intake both of material and of energy. In each step some of the energy is used, less is passed on; the material is all either passed on or returned to air, water, and soil for possible later return by plants. All along the line some organisms die without being eaten. Their bodies and those of the final consumers, the carnivores at the end of the local food chains, decay or decompose, usually with the intervention of bacteria. This process essentially ends the flow of available energy and brings materials to a point where they may again start through the cycle.

That much, at least, must be known to understand the functioning of any living community. Roles of organisms in fossil communities must also be interpreted in the light of these facts. Relationships between the various species of the community are extremely intricate and varied but most of them can be summed up under the general headings of consumption, competition, and cooperation. Of these consumption is perhaps the most complex. It includes, among other things, eating of plants by animals; eating of animals by other animals, including predation; parasitism both by and on plants and animals, which is a special form of consumption without immediate killing of the organism consumed; and bacterial decay, a form of scavenging consumption. Competition, which may be and indeed usually is passive rather than combative, is mainly for sources of energy and materials—for instance, for sunlight among green plants or for prey among carnivores—but may also be for many other needs, such as nesting or hiding places. Cooperation is most importantly involved within the breeding and, in some animals, social structure of a specific local population. It may, however, also involve many other relationships, some of them very curious. A few examples will be given below; here I may just mention the fact that reef corals depend for much of their large oxygen requirement on microscopic plants within the living tissues of the coral animal. This is why reef corals cannot grow below the penetration of light: it is the little plants, not the coral itself, that require light. Nonreef corals do very well in the complete, eternal darkness of the sea's abysses.

All of these relationships are well illustrated by fossil communities, and many of them can be followed by inference through much of the history of life. Simple identification of a green plant forthwith identifies its essential role in the community, although the parceling out of this activity among the various species of plants may be complex and puzzling. Identification of an animal as a herbivore or a carnivore also places it in the major cycle. Further useful analysis requires, at the very least, approximate designation of the sorts or parts of plants and animals eaten. The existence of bacterial decay can and usually must be taken for granted.

It may be of interest first to give a few special examples of interrelationships demonstrated by fossils and then briefly to exemplify the study of fossil communities as a whole.

The predator-prey relationship is especially well illustrated by land-living fossil vertebrates. When such faunas are even moderately well known they almost always include both herbivores and carnivores of appropriate sizes and sorts to prey on the herbivores. The famous *Tyrannosaurus,* one of the largest (although probably not, as often billed, the very largest) of carnivorous dinosaurs, lived among and undoubtedly ate great numbers of herbivorous ceratopsians (*Triceratops* and its allies) and hadrosaurs (amphibious, so-called duck-billed dinosaurs). Eohippus had to contend with a variety of agile carnivores similar to it in size and speed. And so it goes throughout the history.

Usually, as would be expected, fossil herbivores are much more numerous than carnivores. Such a ratio is necessary in a continuing biocenose, because a predator cannot successfully prey on animals greatly larger than itself. Therefore there must be many more prey animals than predators if both are to be maintained. Exceptional thanatocenoses may have many more carnivores than could occur in the biocenose. This is true of the tar pits of Rancho La Brea in Hollywood, where sabertooths (so-called tigers but not really any more tigers than tabbies) are extremely numerous. They succumbed to traps baited with their accustomed prey, ground sloths. The sloths became entangled in the sticky asphalt and sabertooths attacking them became stuck in turn. The extraordinary circumstances resulted in a concentration of remains of predators in a ratio quite impossible in a living community.

Sometimes carnivorous animals are fossilized along with remains of their last meals. The exceptionally preserved ichthyosaur (marine reptile) skeletons from Holzmaden in Germany often have food remains enclosed. They ate mostly squidlike animals (belemnites), ammonites, and fish. A small dinosaur skeleton from the Jurassic of Bavaria has most of the skeleton of a still smaller reptile inside it: the victim was evidently swallowed whole. Similarly, a number of fossil fishes are known with remains of other, food fishes inside and in at least two cases even the skeletons of small land reptiles.

A common form of predation among shellfish today involves sea snails that prey especially on clams and their relatives. The snails bore a hole in the clam shell, then insert a trunklike projection by which the soft parts of the clam are eaten. The telltale holes are common in shells of fossil clams and some other shellfish and testify that this sort of predation has continued through the greater part of molluscan history. Shells, in this case those of brachiopods, pierced by snails are known from the Ordovician.

A sort of inextricable specific community is represented by colonial animals, in which the colony consists of numerous organically continuous individuals developed by budding one from another. The process does not occur among vertebrates, although it does among their relatives the sea squirts or ascidians. It is very common among other animals, those collectively called invertebrates, both recent and fossil. Both solitary and colonial corals have occurred throughout the history of the group. The colonial forms have always been particularly involved in reef building, although solitary corals also occur on reefs. The extinct graptolites, a group of marine animals of unknown affinities, were all colonial. So, with rarest exceptions, were and are the bryozoans, so-called sea mosses or moss animals, which have been abundant since the Ordovician and have also played an important part in reef building.

Also frequent and of peculiar interest are associations in which quite different sorts of animals and plants live together in intimate connection. The relationship may be beneficial to all concerned, or beneficial to one party without mattering much to the other, or beneficial to one and definitely harmful to the other. These are often called symbiosis, commensalism, and parasitism, respectively, but usage is not well established; some ecologists call the first mutualism and apply the term symbiosis to all three. The phenomena intergrade and it may be difficult or impossible for a human to tell whether an association is beneficial and if so to which partner. Of course this doubt applies especially to fossils.

A very common association, beneficial to the worm, at least, and apparently sometimes also to the coral, is that of various worms with colonial reef corals in which the worms build tubes that they occupy. Examples are known in fossils at least as early as the Devonian, and the same sort of association is one of the striking features of living coral reefs. Some students have recently claimed that many of the characteristic Paleozoic bryozoans, including reef builders, were not solely bryozoans as hitherto believed but symbiotic associations of bryozoans and lime-secreting marine plants (various sorts of algae). If confirmed this suggestion will have great significance for our understanding of ancient reefs.

Perhaps the most extraordinary example of symbiosis known in fossils is that of *Kerunia,* a fossil abundant in late Eocene rocks of Egypt. Although quite variable in detailed form, *Kerunia* always has a spiral central part, with various projections like a cock's comb, and two large, nearly symmetrical lateral horns. (Fig. 24.) These strange objects created a flurry of interest after they were found in 1901, and their true nature was established by H. Douvillé in 1905. They consist of marine snail shells which were occupied after the death of the snail by hermit crabs and then overgrown by a colonial relative of the corals, *Hydractinia.* Offhand this interpretation seems a little farfetched, but its brilliance and correctness were entirely confirmed by later discovery of living hermit crabs and hydractinians in symbiosis producing forms almost precisely like those of *Kerunia.*

It seems probable that associations in which both parties benefited or, at least, neither one was seriously harmed sometimes evolved into parasitism, in which one party lives at the expense of the other. The best documented example is that of an association of sea lilies (crinoids, animals related to sea urchins and starfish) and their allies with a variety of snails. The association began at least as early as the Devonian. The snails seem at first simply to have lived on waste from their hosts but eventually to have interfered with normal growth.

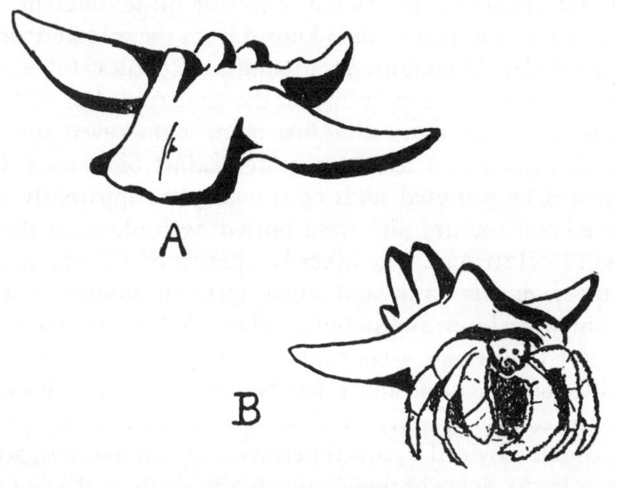

FIG. 24. *Kerunia,* a fossil representing symbiosis between a hermit crab and a coral-like encrusting organism (hydractinian), one living inside and the other outside an abandoned snail shell. A, the fossil as found; the opening of the shell is downward to the left (after E. Fraas, modified). B, schematic restoration with the hermit crab in place; diagrammatic only, no particular species of hermit crab is intended.

Evidences of disease are rather common in fossils and have been studied in great detail, especially by the late R. L. Moodie. They do not, however, suffice for any very significant study of the history of disease. The causative agent in true disease as opposed to injury can never be surely identified in fossils. Furthermore, only diseases that visibly affect the hard parts, shells or bones, are recorded, and these are not, after all, the most important diseases today. It is fairly common for fossil vertebrate skeletons to show abnormal fusion of joints and outgrowths of diseased bone, exemplifying some form of arthritis. Fusion of tail vertebrae in one of the great sauropod dinosaurs is an example, and many other examples are known.

Larger associations, including the whole community or considerable parts of it, are particularly important and are beginning to be studied extensively by paleontologists. By way of examples two instances may be briefly reviewed, both relating mainly to land faunas and fossil vertebrates but otherwise quite different.

A large fauna of mammals is known from the middle Paleocene near the Crazy Mountains in Montana. (Reptiles, fishes, snails, clams, and plants are also known in the same rocks but will not be reviewed here.) An interesting first point is that even among the mammals there are at least two quite distinct biocenoses. One is represented by scattered finds of animals that apparently ranged in more open country and were buried by freshets on the flood plains of local streams. The other is represented by concentrations of bone fragments, jaws, and single teeth of animals that lived mostly in heavy forest and swamps and that fell or were washed into lagoons of the swamp, where they were dismembered by turtles, crocodiles or alligators, and fishes before burial. The flood-plain biocenose consists mainly of relatively large, running animals, about equally divided between herbivorous, omnivorous, and carnivorous forms. Some of these animals also occur in the swamps or quarry biocenose, but there they make up less than half the fauna. Most of this biocenose consists of very small insectivorous or fruit- and seed-eating forms, many of them apparently tree-living, all rare or absent in the flood-plain biocenose. The difference is interestingly exemplified by the fact that two very closely related species of carnivores occur, one species, *Metachriacus punitor,* confined to the swamps and the other, *M. provocator,* to the flood-plain biocenose; the latter is larger.

Compositions of the two faunas are shown in Fig. 25. The following are the general adaptive characters of the major groups shown:

Condylarths: omnivorous to herbivorous preungulates. Larger, more herbivorous forms are common on the flood plain, rare in the swamp. Very small, more omnivorous forms are common in the swamp, rare in the flood plain.

Carnivores: omnivorous to carnivorous primitive forms. Moderate sized to relatively large, more omnivorous, rather bearlike forms are very common on the flood plain, rare in the swamp. Small predaceous forms are about equally common in both biocenoses.

Primates: these were all very small and primitive pre-monkeys. They probably lived largely on fruits, seeds, and the like and were mainly if not entirely arboreal. They are common in the swamp, absent on the flood plain.

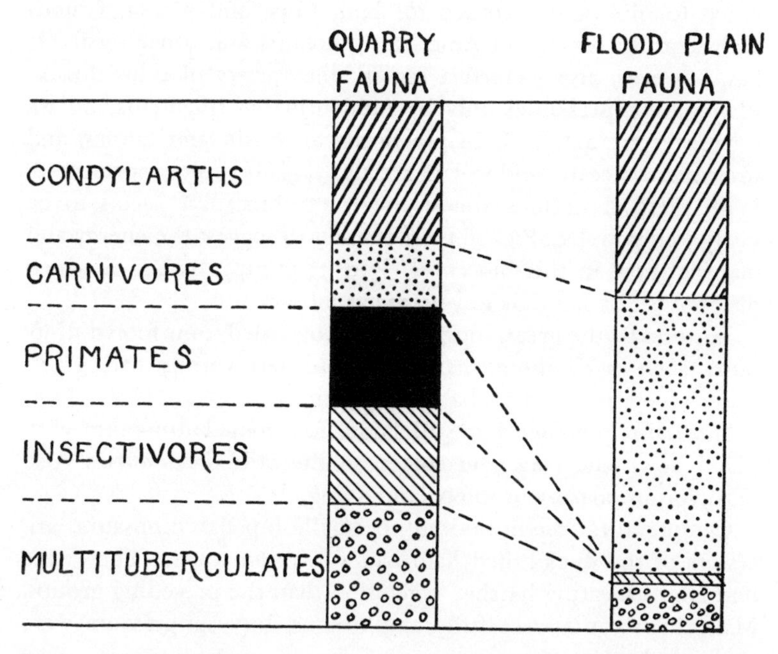

FIG. 25. Fossil facies representing two different biocenoses that lived in the same region at the same time. Proportions of specimens representing various orders of mammals found in a quarry (Gidley Quarry) and in flood-plain deposits in the middle Paleocene near the Crazy Mountains, Montana.

Insectivores: a great variety of small animals. Their habits were evidently quite diverse but many of them probably ate small animal food such as insects, worms, and grubs, and many may have been arboreal, although some lived on the ground and even burrowed. They are common in the swamp, extremely rare on the flood plain.

Multituberculates: an extinct group of small, somewhat rodent-like mammals. They apparently ate seeds and other coated, tough, or fibrous vegetable matter. Some, at least, were probably arboreal. They are abundant in the swamp, uncommon on the flood plain. The flood-plain forms are relatively large and some are more strictly rodent-like.

The second example is based on the fauna of the Morrison formation, late Jurassic, at Como Bluff, Wyoming. This is a locality noted for dinosaurs, worked for both Cope and Marsh, famous 19th-century students of American dinosaurs and other fossils. It has, however, also yielded a remarkable variety of other fossils. Plants are represented only by unidentifiable fragments, microorganisms are absent as in practically all fossil land faunas, and insects, crustaceans, and worms are also vaguely or not represented. With these exceptions almost the whole biocenose seems to be fairly well sampled. Fig. 26 is an attempt to suggest the energy and material flow in this biocenose. Extinct groups, names of which may not yet be familiar to you, are as follows:

Sauropods: the great, long-necked, long-tailed, four-footed dinosaurs, such as *Brontosaurus*. They were herbivorous, eating soft water vegetation, and probably amphibious.

Stegosaurs: smaller but still large, four-footed dinosaurs with plates down the back and spikes on the tail: *Stegosaurus*. Also herbivorous, mainly or solely land-living.

Camptosaurs: medium-sized to small, bipedal dinosaurs, ancestors of the duck-billed hadrosaurs: *Camptosaurus*. They were herbivorous, eating harsher vegetation than the preceding groups. Many hadrosaurs were fully amphibious, but camptosaurs were probably less so.

Carnosaurs: the bipedal, carnivorous dinosaurs, large to small, for example, *Allosaurus* or *Antrodemus*. They were rather varied and apparently adapted to different sorts of prey.

Multituberculates: ancestors of those mentioned above as occurring in the middle Paleocene. Small, somewhat rodent-like, herbivorous mammals.

Pantotheres: also small mammals, but insectivorous in dentition, which means that they probably ate smaller animal food such as worms and grubs.

Triconodonts: another group of mammals. Small in size but apparently fiercely predaceous. They probably killed prey up to their own size or larger.

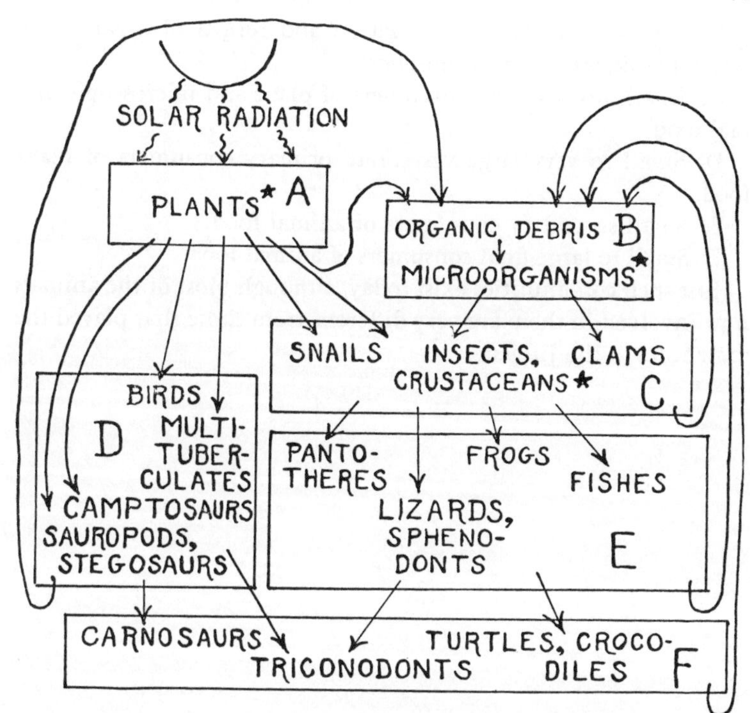

FIG. 26. Partial diagram of food and energy flow in an ancient biocenose. Based on fossils from the late Jurassic (Morrison formation) of Como Bluff, Wyoming. Groups marked with a star are here rare or absent as fossils. Further explained in the text.

Sphenodonts: lizard-like small reptiles related to the living tuatera or *Sphenodon* of New Zealand, which has changed very little since the Jurassic.

This is a large and complex biocenose. It includes aquatic, amphibious, land-dwelling, and even flying animals, but they all lived in one small area and all interacted to some extent.

The chart is very much simplified. Some groups present are omitted, and the various lines of flow were certainly more numerous and complex than shown. Nevertheless it brings out the broader features of community organization. The following general groups and functions are represented:

A. Prime fixers of solar energy and producers of food.

B. Last steps in energy utilization and return of organic materials to soil, water, and atmosphere.

C. Small, invertebrate consumers of plant and microscopic animal food.

D. Small to very large vertebrate primary consumers of plant food.

E. Small secondary consumers of animal food.

F. Small to large final consumers of animal food.

Just such communities exist today, although most of the animals now involved in them are very different from those that played the same roles in the Jurassic.

6. *Fossils and Geography*

ONE OF the really striking things about the life around us today is its geographic distribution. Everyone knows that to see reef corals you must visit subtropical or tropical seas. Kangaroos are found in Australia. Giant redwoods grow only near the Pacific coast of temperate North America. No single sort of life does or can occur everywhere. This is obvious enough from the fact that none can live permanently both in the sea and on dry land. Even if we confine attention to only one of these major environments, no species is ubiquitous. Since conditions in the sea vary less than those on land marine organisms are often very widespread (although some are also very local), but none occurs literally everywhere in the seas. On land restricted distribution is a rule without even near exceptions. The nearest exception is man, with some of his companions and pests, but even man has not yet permanently colonized the vast Antarctic.

Not only individual sorts of organisms but also associations of them, faunas and floras or, together, biotas, are characteristic of different regions and restricted in distribution. The reefs of warm seas involve not only corals but also a great number of other organisms, mollusks, fishes, worms, seaweeds, and many others, that have some degree of similarity everywhere in the reef zone but that form a sort of biota found nowhere else. Not only kangaroos characterize Australia but also bandicoots, koalas, and many other strange mammals, and not only these but also the many species of eucalypts and of wattles, *Casuarina,* and other native trees and scrub.

If we study quite local differences in biotas these are evidently related to living conditions in various places. Different animals and plants occur on prairies, in forests, or among dunes. If in the same general region we go to other prairies, other forests, or other dunes we find essentially the same biotic associations. These differences, in short, are ecological. They are facies such as have previously been discussed in this book and are not in any other or special sense geographical. If, however, we move to quite a different region another factor clearly comes into play. We may find prairies almost exactly like those at home in appearance, elevation, and general living conditions. Then the ecology, too, will be found quite similar. The biocenose has the same possible ways of life or roles and these are played in similar ways. But they are likely to be played by quite different animals and plants. Herds of animals crop bushes in Europe, Australia, Africa, and South America, but the animals and the bushes are not the same in any two of these continents. Here are differences that have some deeper geographical basis and that are not only ecological.

In the late years of the 19th century naturalists (among them the Sclaters, Beddard, and Lydekker) distinguished a few major faunal types among recent land animals and divided the world into corresponding faunal realms or regions. The system was based on land birds and mammals, but with modifications it also applied fairly well to some other animals and even to some groups of plants. Different names were applied and there were differences of detail as to boundaries, but one form of the agreed arrangement is as follows: Africa north, roughly, of the Sahara, all of Europe, Asia north of the Himalayas, and North America north of the tropics form a single major region, the Holarctic. This vast area has surprising similarities of fauna from end to end. It may, however, be subdivided into Old and New World parts, Palaearctic and Nearctic, respectively. Various other subdivisions can also be recognized, such as the Mediterranean in the Palaearctic and the Sonoran in the Nearctic. Africa south of the Sahara is another major region, the Aethiopian. Southern Asia and most of the East Indies are the Oriental Region; Australia and adjacent islands the Australian Region. Finally, South America, tropical North (or Middle) America, and the West Indies comprise the Neotropical Region. (Some of these names are not very well chosen: there is nothing especially new about the "neotropics" and the Neotropical Region is by no means all tropical.)

Such an arrangement is a reasonable description of certain facts about living animals. The division lines are not nearly as sharp in nature as they are on the map and many details are disputable, but by and large there are sorts of faunal associations corresponding with these geographic regions. The picture is, however, a static one, representing only one moment in the long history of life. It explains nothing and is even rather meaningless unless it is considered as the result of a series of historical events. Why, for instance, should the Neotropical and Nearctic faunas be so different when there is no real barrier between them? On the other hand why should the Nearctic and Palaearctic parts of the Holarctic fauna be so similar when there *is* a decided barrier between them—the Arctic Ocean and its connections with the Atlantic and Pacific?

When the designated faunal regions are viewed historically, with the aid of fossils, it appears that they may hinder more than they help understanding. At least, they tend to put things in the wrong terms and to analyze the wrong elements. The llama is a typical Neotropical mammal, but not long since, geologically speaking (in the Pliocene), it was entirely absent in the Neotropical Region and its forebears were typical of the Nearctic Region. On the other hand, other Neotropical animals, such as the armadillos, were then more exclusively Neotropical than they are today. Or, again, note that Middle America now has a somewhat special but still distinctly Neotropical fauna as a whole, and that in the Pliocene it had a completely North American fauna with no resemblance to the fauna of South America.

Evidently the Neotropical fauna has not long had the same composition as it has today and has not long had its present distribution. In fact when you really try to understand the fauna and to analyze it in an explanatory way, the term and concept "Neotropical" simply become confusing. Much the same is true of any analysis in terms of recent faunas and faunal regions.

It is more enlightening to think of the land surface of the world as made up of a number of blocks, great segments of the earth's crust. The major blocks correspond to the present continents: Eurasia, Africa, Australia, North America, and South America. (Probably Antarctica should be included, although its part in the history of land life is dubious, and it now has virtually no land life.) It is rather generally agreed today that these blocks have retained their identities throughout the history of land life. They have, certainly, changed greatly in outline and detail, and whether they have maintained approximately their present positions is a disputed point mentioned later in this chapter. Neither of these factors negates continuous existence of the blocks as such.

The blocks have sometimes been connected with each other in various ways and sometimes each has been isolated from all others.

Africa has been connected to Europe and to Asia, as it is, tenuously, at present. Europe and Asia have been so often and broadly connected, as they are now, that they really form one very large block. Australia has been connected to Asia. North America has been connected surely to the Asiatic end of Eurasia, probably also to the European end. South America has been connected to North America, as at present and more widely. Some students have postulated other connections, such as between Australia, South America, and Africa, but the connections noted above are the only ones that can be called well established. (My personal opinion is that if any others ever occurred it was so long ago that it had no perceptible influence on present faunas and little or none on the history of land life unless near its very beginning.)

Whenever a connection existed forms of land life spread from one block to another. When a connection did not exist the intervening sea was a powerful barrier to such spread. Land organisms, especially plants but also some animals, have occasionally spread across the sea barriers, but extensive interchange of faunas requires a land connection.

The whole history has been greatly complicated by changes within each block. Ecological subdivision has, of course, always existed and has made for restricted and uneven distribution of each sort of organism on any one block. In addition to this intricate and shifting factor geographic barriers have occurred within as well as between blocks. Two major barriers of this sort exist now and have for some time past: the Sahara (and associated deserts), a zone of aridity across Africa, impassable for many animals and plants; and the Himalayas (and associated uplifts), a zone of rugged, cold country across Asia, also impassable for many animals and plants. The effect of such barriers is especially strong when, as in these two cases, they run from east to west, in the same direction as climatic zones.

In the past there have also been sea barriers within as well as between the continental blocks. Relatively shallow seas have repeatedly flooded right across the blocks and cut them into two or more separate land areas. These seas certainly were barriers to land life and affected its history, but they fluctuated and usually did not persist as complete barriers for long periods, geologically speaking. Moreover, they apparently have not completely isolated two major segments of a single block for significantly long periods of time since early in the Cenozoic, at latest, and so have not had a really strong influence on distribution of most living faunas and floras. These complicating factors are important and have modified many details worthy of (and receiving) study. On the whole, however, most of the major features of the history of land organisms can be followed sufficiently well by reference mainly to the continental blocks as units and to their connections and disconnections.

One of the most completely analyzed examples of intercontinental faunal relationships is that of the land mammals of Eurasia and North America from early Eocene to Recent. Simultaneous appearances of closely similar animals on the two continents show that migration has occurred between the two. By counting the numbers of such occurrences at different times, the varying extent of faunal exchange can be measured in approximate and relative terms (see Fig. 27). For various reasons there is some inaccuracy and lag in such measurements, but the data clearly show that there was extensive interchange and therefore almost certainly a land connection between the continents during most of the Cenozoic. Interchange was especially strong in the early and late Eocene, early Oligocene, late Miocene, and Pleistocene. It was particularly weak in the middle Eocene, middle to late Oligocene, and Recent. The connection is known to be broken now and it doubtless was also at the earlier times mentioned. A briefer interruption sometime during the early Pliocene is also probable.

Resemblance in faunas of the two continents was of course increased by each extensive faunal interchange. Resemblance is, however, affected by several other factors as well and does not simply reflect interchange or making and breaking of the connection between the continents. For instance, the extinction of groups of animals typical of one continent and not the other increases the resemblance because it decreases the difference between their faunas. So it happens that at present, with no interchange going on but following extensive Pleistocene interchange, the resemblance between the mammals of Eurasia and those of North America is very high. Measurement of resemblance is of course based on the numbers of groups occurring in both of the two regions and those present in only one of them. There are some tricky points in the statistics of getting a really valid comparative measurement, but we need not go into those here. Resemblance can be measured and followed through time (Fig. 27). You will notice that faunal resemblance between Eurasia and North America has been much lower at all times since the early Eocene than it is at present. It was especially low in the middle to late Eocene and late Oligocene to early Miocene. At those times, at least, the concept of a Holarctic Region is not applicable. In fact it hardly seems to apply to any times except the early Eocene and the present.

In such cases study of local faunas within continents and of the sorts of animals that did or did not spread from one to another may give strong evidence as to where the land connection was and what it was like. In the case of North America and Eurasia, through all the history since middle Eocene the North American faunas are most like those of northeastern Asia. (Earlier faunas are not well enough known in Asia to make the comparison.) There is also evidence that as a broad average the animals that made the trek could tolerate cooler climates and more northern conditions than those that did not spread from one continent to the other. The connection seems surely to have been between eastern Siberia and Alaska. During this time, at least, a direct connection between North America and Europe is ruled out. A final interesting point about this example is that animals spread in both directions across the land connection, as would be expected, but that rather more spread from Eurasia to North America than in the reverse direction.

Such a history is extremely complex in detail. It is compounded of the separate histories of many different groups of organisms. Some spread in one direction, some in another, and some in neither. Within each region groups tend to change in different ways. In each region some groups become extinct. Moreover, the whole story is not simply one of exchange between two regions. There are movements and changes among and within parts of each region, and each may periodically exchange groups of animals with still other regions. The composition of the fauna is thus in a constant state of flux. At any one time the fauna is an extremely heterogeneous mixture—and yet one that is adjusted into an ecological community—of animals the ancestors of which came at different times from different places and which have changed in different ways and to different degrees since their ancestors arrived.

Such a complex fauna (and almost all faunas have this character

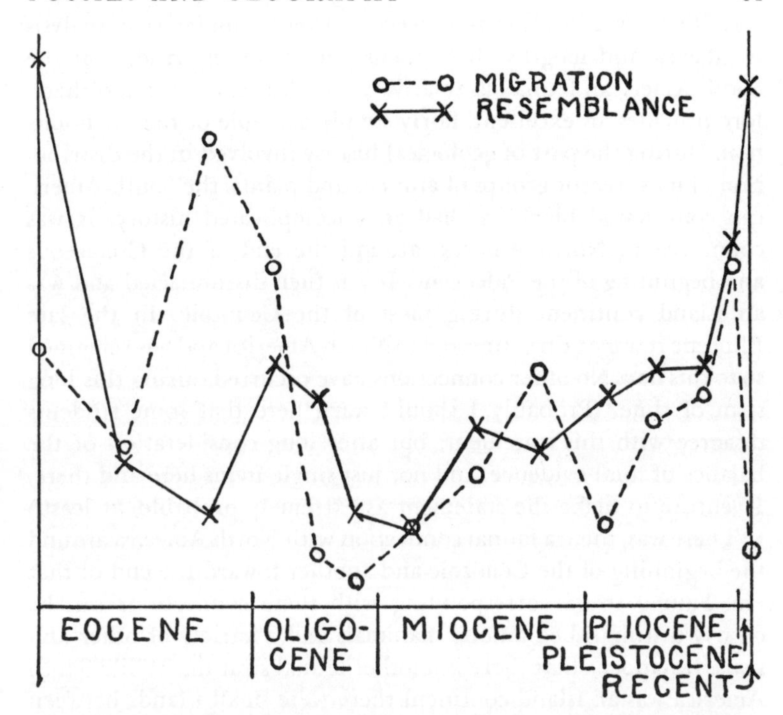

FIG. 27. Amount of intermigration of land mammals and degree of resemblance in faunas between Eurasia and North America during the last 60 million years. The horizontal scale is proportional to estimated lapse of time in years. Crosses and circles are calculated values for each third of an epoch from Eocene to Pliocene and for the Pleistocene and Recent.

to some degree) may be said to be stratified. The oldest stratum consists of animals that have been present in the region for the longest time. On an average they will have changed most since their ancestors entered the region. As a result they will be least like the animals of other regions, unless they have themselves spread to those regions. Newcomers, on the other hand, will have changed least since arrival. They will still be closely similar to relatives in the region from which they came, unless those relatives have become extinct in the meantime.

The European and North American faunas are stratified in this way. Their stratification is, however, so very complex that analysis is difficult and lengthy. It happens, on the other hand, that the South American fauna has relatively few clear-cut strata and therefore provides an excellent, fairly simple example of the phenomenon. During the part of geological history involved in the distribution of most recent groups of animals and plants, the South American continental block has had an uncomplicated history. It was connected to North America around the end of the Cretaceous and beginning of the Paleocene. It was then disconnected and was an island continent during most of the Cenozoic. In the late Pliocene it was again connected to North America and has remained so to this day. No other connections have occurred during this long span of time. (Probably I should warn here that some students disagree with this statement, but after long consideration of the balance of total evidence, and not just single items here and there, I venture to make the statement as extremely probable, at least.)

There was, then, a faunal connection with North America around the beginning of the Cenozoic and another toward the end of that era. Faunal strata corresponding with these connections can be clearly distinguished among mammals and a variety of other animals. It happens that there is another faunal stratum. While South America was an island continent there were small islands between it and North America, where Middle America now stands. Across this chain two sorts of animals managed to island-hop to the southern continent: primitive rodents and monkeys. They did not do so at exactly the same time. The rodents probably made the hop in the late Eocene and the monkeys sometime around middle Oligocene. The last faunal interchange was also quite complex, starting with island-hopping and with several later waves of land migration.

In spite of these complications it is broadly evident that South American mammals and some other animals are separable into three main faunal strata: the oldtimers, whose ancestors entered South America around the beginning of the Cenozoic; the older island-hoppers, whose ancestors reached there in late Eocene or Oligocene; and the newcomers, which began to appear in latest Miocene but most of which came in the latest Pliocene and subsequently.

These strata correspond with degrees of distinctiveness of the animals involved. The armadillos or tree sloths, surviving old-timers, are mammals (of the type called "placental" in contrast with marsupials), but aside from that they resemble nothing else on earth and have no special relatives in any other region. The South American monkeys, surviving older island-hoppers, clearly do have relatives elsewhere, the African and Asiatic monkeys. Nevertheless they are quite distinct as a group from the Old World monkeys and not as closely related to them as you might suppose from the loose application to both of the same vernacular name "monkey." The native rats and mice, deer, wild dogs, and numerous other newcomers are still very like their North American relatives. (Fig. 28.)

The earlier faunal strata included many groups of animals that became extinct, mostly because they were displaced by members of the later strata. For instance, development of the oldtimers included a great variety of hoofed herbivores. Some of these, the smaller, rather rodent-like ones, were displaced by true rodents as the second faunal stratum evolved. The rest were displaced by hoofed herbivores among the newcomers (horses, tapirs, peccaries, camels, deer), and none survive today. Many of the older native rodents, in the second faunal stratum, also became extinct when the rabbits and ratlike rodents of the third stratum came in. The present "Neotropical fauna" is made up of nearly half late immigrants, really belonging to "Nearctic" groups, over a third old island-hoppers, and only a sixth oldtimers, the only groups that are fully "Neotropical" on the basis of long differentiation in that region and absence of close relatives evolved in other regions. (Fig. 29.)

In Middle America similar faunal strata occur, but their geographical history there is quite different, indeed practically reversed. The oldtimers of South America are the newcomers in Middle America and so, probably, are the South American older island-hoppers. In Middle America the forms that are newcomers in South America belong to a complex sequence of older faunal strata related to those of North America and not present as separate strata in South America. The West Indies, also Neotropical to earlier zoological geographers, have had a faunal history radically different from that of either Middle or South America and have a faunal stratification unique to them. Without further detail you can see that such a term concept as "Neotropical" simply does not make sense when you try to understand what really has gone on and what the geographic relationships of faunas really are.

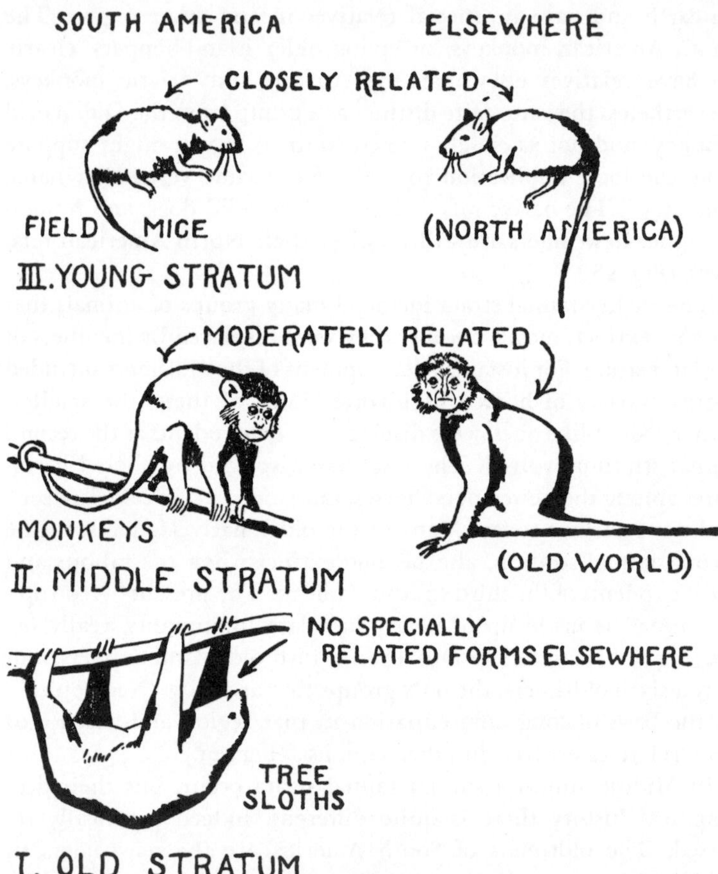

FIG. 28. Living representatives of the three major faunal strata of South America and their closest relatives elsewhere.

Implicit in all such faunal histories are differences in the ways in which organisms spread from one place to another, that is, in their means of dispersal. For instance, some land organisms practically require dry land for their dispersal. Elephants are an example, but even elephants have been known to cross rather wide stretches of sea between islands. The requirement is never 100%.

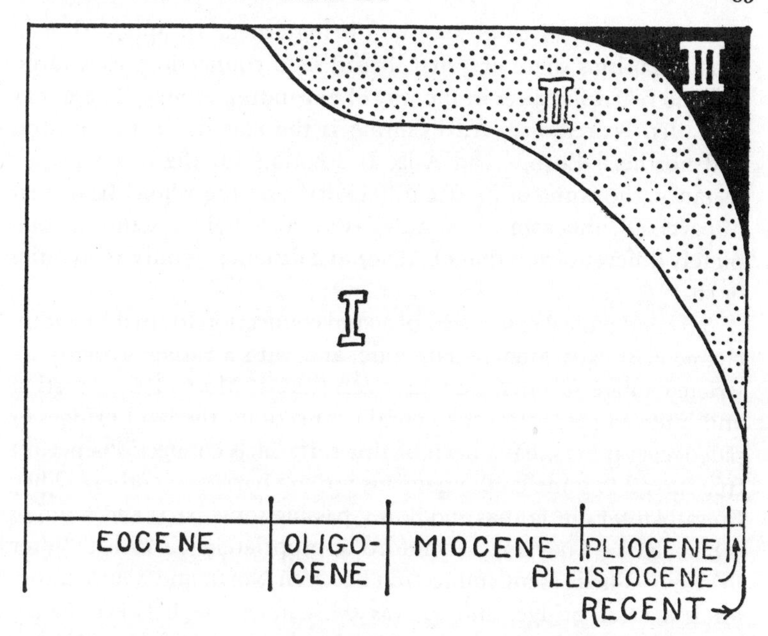

FIG. 29. Composition of the South American mammalian fauna during the last 60 million years. The diagram shows the percentage of known forms belonging to each of the three main faunal strata. The horizontal scale is proportional to estimated lengths of the epochs in years.

It is merely extremely improbable, not absolutely impossible, for such animals to spread without a land connection. (That is a point that many students of plant and animal geography have overlooked.) For other land organisms water is hardly a barrier at all. This is notably true of plants with seeds dispersed by the wind. Land snails on the Pacific islands have spread freely over tremendous stretches of ocean, although no one is yet quite sure how they managed to do so. Organisms to which water is a serious barrier may nevertheless manage to cross if enough time elapses—and geological history allows a fair supply of time in most cases.

Along this scale there is continuity and no hard and fast distinctions between, at one end of the scale, barriers that make dispersal *almost* impossible and, at the other, migration routes that make spread *almost* certain. For purposes of descriptive analysis, however, three typical cases may be distinguished. In application to land organisms there are, first, broad land connections that interpose no real difficulties in the way of expanding groups. These may be called "corridors." An example is the corridor across central and northern Europe and Asia. It accounts for the fact that, for instance, the fauna of France differs little, on the whole, from that of parts of China some 5000 miles away, although it is almost completely different from that of Africa at a distance of only 1000 miles or so.

A second typical case is that of a land connection limited in width, characteristically longer than wide, and with a rather strongly restricted range of environmental conditions. Most of the existing and inferred past intercontinental connections, the land bridges of paleogeographers, have been of this sort. Such connections permit free spread to some land organisms but are barriers to others. They literally filter out faunas and floras, passing some parts and holding others back. They may therefore appropriately be called "filter bridges." The present connection between North and South America is a filter bridge, and so was the connection between North America and Asia. Many plants and animals spread across it but never the whole or even much the greater part of a flora and fauna. It was always selective.

Finally, there is the case of a strong barrier which nevertheless is occasionally passed by one group or another. For land organisms this is often exemplified by a wide stretch of sea with scattered islands. Because of the large element of chance, with odds all against success, I have called such avenues of rare dispersal "sweepstakes routes." We saw, above, that rodents and primates reached South America by a sweepstakes route. It is (in my opinion, which some students do not share) to sweepstakes dispersal from Asia that Australia owes the peculiarity of its fauna rather than to oddities of times and places of former land connections.

A parenthetical remark may be inserted here. Discussion in this chapter has been based on land organisms and especially on mammals. This is because there are available particularly good and extensively analyzed examples involving mammals and because I happen personally to know more about these examples. With appropriate allowance for differences in means of dispersal, character of barriers, and the like, the same methods of analysis and the same concepts can be applied equally well to all sorts of organisms, land plants, marine mollusks, or any others.

Students of the geography of living animals and plants have been particularly fascinated by what they call "disjunctive distributions." These are cases of the occurrence of the same or closely related organisms in widely separated areas, with no equally related forms anywhere between them. Famous examples are the presence of tapirs in southeastern Asia, Middle and South America, and nowhere else; and the occurrence of wild araucarias, a distinctive conifer widely cultivated under a variety of names, in Australia (also Philippines to New Zealand) and South America but not in Africa or the northern continents.

It used to be a popular pastime to postulate an ancient land bridge to account for each separate case of disjunctive distribution. Compilations of a generation or so ago show the Atlantic, Pacific, and Indian Oceans crisscrossed by land bridges throughout most of geologic time—land bridges most and perhaps all of which never existed outside of the imaginations of their proposers. It cannot be said that modern students have entirely abandoned the reckless, nonscientific proposal of such bridges—I have already had occasion to repeat that old errors never die—but most of us have learned to evaluate evidence more rationally.

In most cases when related fossils have been found it has turned out that groups now disjunctive once occurred continuously across land still existing between their recent areas. Tapirs, for instance, were formerly abundant in North America and right across Asia into Europe. They have simply become extinct except in two of the southermost parts of their former range. Araucarias, too, were formerly widespread throughout the northern hemisphere. (The most abundant trees in the famous petrified forest of Arizona are ancient araucarias.) Their distribution requires and suggests no land bridges across what are now ocean basins. Some students of groups for which fossils are unknown or nearly so (worms, for example) still insist on accounting for disjunctive distributions by land connections across the present oceans. Surely it is more reasonable to suppose that the explanation is the same as for the groups in which adequate fossil evidence is available.

Another point is that in the vast stretches of geological time sweepstakes dispersal has been much more common than many students have realized. Some organisms have in fact crossed the oceans, but not on land bridges. This doubtless explains some of the similarities between the biotas (especially the floras) of Africa and South America, although some are explicable, as above, by dispersal through the northern lands. There still seems to be a consensus among well-informed and reasonable students that a number of plants (but few or no animals) spread between Australia and South America without crossing the Pacific or the northern continents. The most likely hypothesis is that there were sweepstakes routes (but not land connections) from Antarctica to Australia and to South America and that this was the line of dispersal for some plants when the Antarctic climate was milder than at present.

It seems probable that all major features of animal and plant geography can be and should be accounted for without assuming any great past differences in positions and relationships of the continents. The only relatively recent (that is, say, Cenozoic) major intercontinental connection not now in existence but really supported by good evidence is that between Asia and North America. There were probably earlier connections between Australia and Asia and between Europe and North America, but both were too ancient to have much bearing on the greater parts of the recent faunas and floras of those continents. Other extremely ancient (Paleozoic or earlier) connections may have existed, but evidence for them is inconclusive and they are not likely to have any bearing at all on the distribution of modern biotas.

This conclusion is of course a personal opinion, but it is an opinion based on many years of careful study and as dispassionate a review as possible of all opposing opinions. The most strongly opposing opinion still held by some competent students involves the theory of continental drift. There are several contradictory versions of that theory, but they all suppose that the continental blocks formerly had quite different relationships and positions on the spheroid of the earth. They are supposed to have formed either one or two land masses which subsequently broke up into the existing blocks and then drifted apart and finally into their present positions. Different adherents of the theory have markedly different views as to the time and sequence of the breaking apart of the blocks and the rate and direction of their drifts to their current places.

There is a considerable amount of strictly geological evidence that can be interpreted as suggesting that some such process may sometime have occurred. The geologists are very far from agreeing not only as to details of the process but also as to whether it did in fact occur. Most American geologists believe that it did not. European geologists are more divided in opinion, and more of them are inclined to accept the probability of continental drift. South African geologists are especially favorable to the theory, largely because of the personal influence of an able and inspiring exponent of it, Alex. L. Du Toit, who worked and taught in South Africa.

Besides the strictly geological evidence, discussion of which would be out of place in this book, adherents of the theory have all related it to the geographical distribution of plants and animals. The discussion, much of it by geologists not really familiar with plants and animals, has contained many errors of facts and of logic in this field. Disregarding these failings, I think it fair to say that what is really known about past and present plant and animal geography lends no adequate support to the theory of continental drift. Any influence of drift on biotic distribution from about the Jurassic on is rather conclusively opposed by the balance of evidence. For the Triassic and earlier periods the possibility is not conclusively excluded, nor yet really supported, by the fossil evidence. Because Triassic and earlier distributions have left little effect on modern distributions, continental drift, even if it really occurred, could not help significantly in explaining the present geography of plants and animals.

The topics that have been discussed in this chapter are special aspects of paleogeography, the study of changes in geography through geological time. One aim of that study is the production of maps of the world as it has been at various periods in the past. The most obvious feature of such maps must of course be the contrast of land and sea, requiring the tracing of past coastlines. This can generally be done approximately and sometimes in full detail when the ancient coast occurred where now is dry land. Then the trace of the coast is often preserved in rocks where it can be studied.

But ancient coastlines also commonly occurred in places now covered by the sea and not accessible for adequate study. (A science of submarine geology has recently been developed and is making remarkable strides, but it has not yet significantly modified this restriction and is not likely to for a long time, at least.) In large part, the evidence, such as it is, for such coastlines has been derived from studies of past and present distributions of organisms as indicative of former land and sea connections. Many paleogeographic maps of the world were produced on this basis, but it must now reluctantly be decided that few of them were worth the paper they were printed on. This gloomy conclusion is sufficiently supported by comparing some of the maps published by different men of equal learning and skill and purportedly showing the earth at the same time. Often they do not look as if their authors had been working on the same planet.

One trouble was that paleontologists differed so radically as to where and when land and sea connections occurred, and many of them were recklessly building land bridges (in their own minds, only), as noted above. These difficulties can be straightened out, and indeed are being well cleared up as more facts are learned and more logical interpretations achieved. Another difficulty is likely to be permanent. Studies like those of the Eurasian–North American faunal interchange can reveal when an intercontinental connection occurred and even approximately where it was. They cannot possibly show its outline, position, width, or other geographic features clearly enough to call any drawing of it a scientific *map*. The result can only be a diagram—or a flight of fancy. Such diagrams are, indeed, useful and significant scientific results if they are plainly diagrams and do not masquerade as maps. (Fig. 30.)

These difficulties do not apply to paleogeographic maps of limited regions now accessible for detailed geological study. In this more restricted field paleogeography has become a science fully worthy of the name. The geologist (that variety called "stratigrapher") and the paleontologist, in one person or working together, are achieving quite remarkable results in following local and even, with decreasing precision, continental geography through past periods. The methods are more geological than paleontological, but they require aid from the paleontologist. One such map

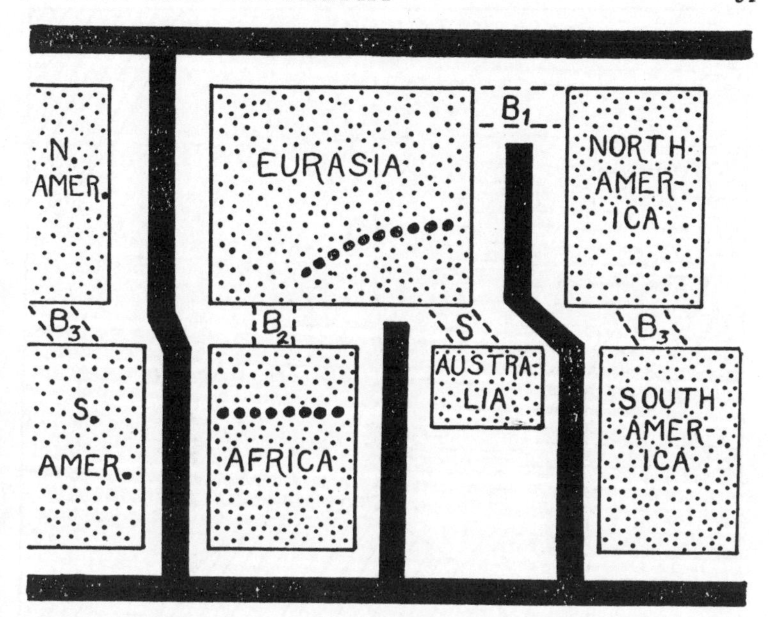

FIG. 30. Diagram of major features of the geography of the world bearing on present distribution of land animals. This diagram is believed to sum up such major features for the past 50 million years, at least, and probably 75 to 100 million years. Heavy black lines represent sea barriers permanent during that time. Heavy black dots indicate two major land barriers arising during that time and becoming progressively more important. B, the three intercontinental land bridges, variable during the time in question. S, a variable sweepstakes route.

is given by way of example (Fig. 31). If you are particularly interested in this sort of reconstruction of the past you will find discussions and other examples in recent textbooks of historical geology and stratigraphy, some of which are listed at the end of this book.

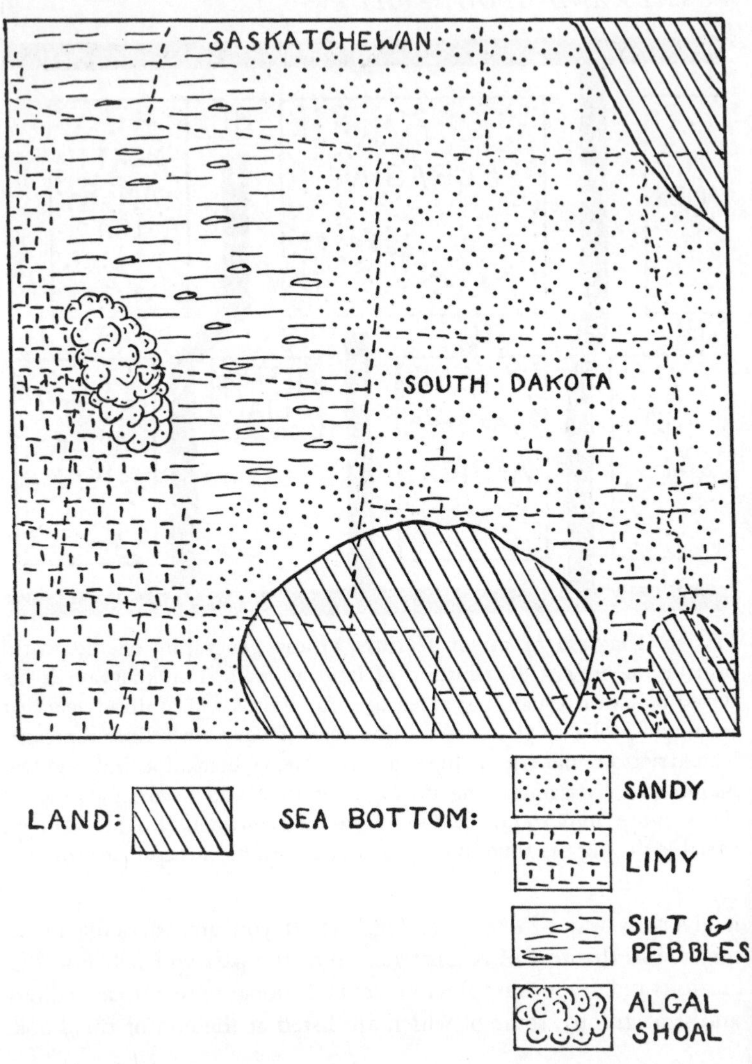

FIG. 31. Geography of part of the interior of North America during a stage (Franconian) of the late Cambrian. The ancient geography is superposed on a sketch map of part of the United States and Canada with outlines of the states and provinces, one of each of which is named for orientation. (After Lochman.)

7. *The Diversity of Life*

No one is so unobservant that he has failed to notice and to marvel at the great variety of forms of life existing around us today. It is not necessary to be a zoologist or a botanist to know dozens or hundreds of different kinds of animals and plants. A short walk in forest or field with attention on the diversity of life will surely reveal at least a hundred sorts of living things, from trees to insects, recognizably different even though no one person is likely to know the names of every one of them. This is only the beginning, for there is also secret life all around you: nocturnal creatures hiding until the day ends, burrowers that seldom appear in open air, swarms of microscopic organisms everywhere below the power of the eye to see. Then add all the creatures of other environments and other lands, and those multiplying in the vast oceans. No one has ever counted how many kinds, species, of organisms now exist, but the *lowest* recent estimate for animals alone is about one million. Between animals and plants there are probably several millions of living species. Extinct species must, in sum, have been far more numerous than those now living. The total for all organisms that have ever lived staggers the imagination.

Diversity is one of the great facts of life. It is the delight and also the despair of everyone who studies life, recent or ancient. No rational approach to such study can be made without some way of bringing order into the seeming chaos of life's exuberance. In other words, in dealing with life it is an intellectual and also a practical necessity to classify its forms. The earliest men must have noticed that some plants and animals are more alike than others and have tended to group them by their similarities. Certainly modern uncivilized tribes do this, sometimes in elaborate systems rivaling that of the professional biologist. Reflections of such an untutored classification occur in the Bible. Aristotle and other primitive scientists built up more self-conscious classifications.

With the revival of independent inquiry as the oppression of dogma and authority began to lift with the Renaissance, renewed and increasing attention was paid to this problem. A method slowly worked out was to draw up a sequence of brief descriptions.

The first would be very broad, applying equally well to a great number of plants or animals, for instance, "Plants with flowers," "Animals with blood." Succeeding descriptions would be more and more restricted, finally winding up with an abbreviated, almost code phrase of two, three, or a few words designating just one particular variety of plant or animal. The development of a classification in this way culminated with the Swedish naturalist Linnaeus (1707–78), who apparently invented no part of the system but who put it in consistent shape and popularized it. Superficially, at least, we still use that system and still call it "Linnaean," although most of us understand it quite differently from Linnaeus and his colleagues.

The broader and narrower descriptions correspond with groups or categories including more or fewer sorts of organisms. Although the series from large to small groups really comprises a very large number of possible steps, for practical purposes it is arbitrarily divided into a moderate number which comprise the Linnaean hierarchy. As now used by most biologists the basic steps in the hierarchy from larger, or higher, to smaller, or lower are:

KINGDOM Example: Kingdom Animalia, all animals
PHYLUM Example: Phylum Chordata, animals mostly with backbones
CLASS Example: Class Mammalia, mammals
ORDER Example: Order Primates, monkeys, apes, man, etc.
FAMILY Example: Family Hominidae, humans and near humans
GENUS Example: Genus *Homo,* humans
SPECIES Example: Species *Homo sapiens,* the one living species of mankind

Classification of extremely varied groups now commonly requires more steps, and these are supplied by prefixing "super-," "sub-," or "infra-," to the names of the basic Linnaean categories. Sometimes other categories such as "cohort" or "tribe" are inserted, but there is no general agreement as to their use or place in the hierarchy. The basic hierarchy, as given above, is the one other thing (besides the geological time table, Chapter 3) that *must* be memorized by anyone who wants to read this book, or any other book on zoology, botany, paleontology, or historical geology, and to follow it intelligently. Memorizing seven words and their sequence is no great task, and it is the key to a whole realm of knowledge.

Names applied to a particular group of organisms have been coined by students of those organisms. The names may be of any origin, but they are treated as if they were Latin—a useful relic of the days when Latin was the universal language. From genus upward each name is a single word, always capitalized when used in a technical sense. The name of a species is two words, first the name of its genus and then its own name, not capitalized. Names of genera and species are underlined in manuscript and italicized in printing; names of higher categories are not. Subspecies, smallest category properly recognized in the system, add another italicized name to that of the species.

Many names have appropriate meanings: the Chordata have a noto*chord,* a stiff rod down the back, in early development, at least; the Mammalia have *mammae,* milk glands; man considers his order, the Primates, *prime* among all; and man also likes to think that *Homo sapiens* is *sapient.* It is nice when the names are appropriate and helps to remember them, but there are not enough appropriate names to go around, and a name appropriate for some members of a group may not be for all. (There are strictly herbivorous Carnivora.) Also, the system would fall back into chaos if a name were changed every time someone thought it inappropriate. (*Basilosaurus,* which means "king of the reptiles," is not a reptile but a fossil mammal.) The names are just names and it does not really matter what they mean as words. What they really mean is a particular group of animals or plants.

Later classifiers, the taxonomists or systematists, have taken over the Linnaean hierarchy and still use it for all organisms, living or extinct. There have, however, been two quite revolutionary changes in our understanding of it and in the way in which we actually go about classification put in this form. Linnaeus, his contemporaries, practically all of his predecessors, and many of his successors were simply classifying animals and plants by their similarities and differences, by the number and sorts of characters common to all members of a group and distinguishing it from other groups. Some of them did not attempt to see any meaning in the characters-in-common. More of them thought that the characters-in-common constituted a sort of pattern followed by the Creator, a pattern sometimes called the "archetype."

The first revolution followed general acceptance of the truth of evolution. It then appeared that the explanation of the characters-in-common is that they are inherited from an ancestor-in-common; they indicate material relationships among the organisms concerned. The different categories were then interpreted in terms of descent. All the species of a genus were supposed to have evolved from one ancestry, as were all the genera of a family, all the families of an order, and so on upward. Actually there are a great many complications that do not appear on the surface of this simple statement, but in essence that is the way everyone now understands and attempts, at least, to use the hierarchy. After this revolution, the hierarchy *looked* the same as before, but it *meant* something totally different.

Yet for a long time the actual procedure of identifying and classifying organisms remained practically the same as in Linnaeus' days and earlier, and this is still true today to some extent. Pre-evolutionary students set up an abstract archetype, which involved characters-in-common, and then decided whether a given organism, any individual specimen, belonged in the group by judgment that it was, or was not, enough like the archetype. All early postevolutionary students, and many down to our own day, did exactly the same thing. Each species had a type specimen, and other specimens were identified by comparing them one by one with the type and deciding, by individual taste, whether they were enough like it to belong to that species. Higher categories, genera, families, etc., had "diagnostic characters," which are archetypes although that old-fashioned word is not used now. The synonymous new-fashioned word is "morphotype." The approach to classification by individual comparisons with type specimens and morphotypes is still the typological procedure of pre-evolutionary days.

The second revolution is one substituting truly evolutionary and biologically realistic concepts and procedures for the nonevolutionary and idealistic concepts and procedures of typology. This revolution is more subtle and its significance seems a little harder to grasp, but it more profoundly affects actual classification than did the result of the first revolution, which established that evolutionary relationships underlie the Linnaean hierarchy. The second revolution is still going on. Some students still do not grasp its meaning, and a few seem blithely unaware of it or are simply incapable of changing habits of thought. It is, nevertheless, already so widely understood and so generally followed that there is no doubt that this is the way classification will be understood in the future. This would not matter particularly if it were just a matter of classifying organisms in one way or another, but it is much more than that. It profoundly affects our whole understanding of the world of life. It has repercussions in widely differing views as to how evolution occurs. Ultimately it affects man's understanding of his own destiny and his place in the universe.

The new concepts arise from the facts, which are simple and even seem commonplace nowadays, that organisms vary and that they change in time. Subtleties become involved with realization that it is *populations* of organisms, and not individuals, that vary significantly and that evolve in the course of time. The biological and evolutionary unit in nature is a group of more or less similar organisms living together and, if reproduction is sexual as it is in most organisms of all sorts, interbreeding with each other. The systematist is, or we now know that he should be, classifying populations, not specimens. A population has no "type." It is what it is, a group of organisms, no two quite alike, embracing certain ranges and sorts of variation, and with every variant just as much a member of the population and just as representative as any other. Designation of one individual as a "type" and comparing others with it, each as an isolated object, simply does not make sense in view of these fundamental truths about organisms in nature.

(Even in modern systematics, types are still designated for species, but they are not standards of comparison. They are needed for certain procedures of nomenclature into which it is unnecessary to go here. They are purely arbitrary, legalistic devices, a necessary evil to the professional systematist, but entirely without biological significance.)

In the practice of systematics this concept means that we are not really primarily engaged in classifying the specimens we collect. They are samples from a natural population. We use them as a basis for judging what that population is like, and then we attempt to characterize the population as a whole. The population is what is classified. One specimen tells something about its population. More specimens tell more. The whole sample available, which is technically called a hypodigm, must be studied together, and no specimen must be given more weight than another or considered in any way more "typical" than another. Otherwise judgment of the population will be falsified.

There are complications in gathering good samples and estimating their reliability. Methods of inference as to the populations are also numerous and complicated. This is a field of research that is now very active, changing and becoming more reliable every day. Like anything else in the common-sense procedures of science, the theory and practice of samples can be explained in simple terms and easily learned by any normal person willing to take the time. But the subject is complex and so it does take time. For purposes of this book the only important thing at this point is the concept that specimens are samples and that the things classified are populations in nature. Further, and as a corollary to that, there is no such thing in nature as an archetype or morphotype, and those terms and concepts merely tend to make us falsify or overlook the realities of nature.

These realities do not, then, start with a series of broader or narrower ideal patterns from which imperfect, real individuals differ more or less. They start with groups of individuals which stand in a certain relationship to each other and with a material pattern of diversity in that population. The basic unit of classification, the species, is such a population, or it is a series of such populations incompletely isolated from each other. The populations within a species do not have definitely separate evolutionary roles. Their variation runs through the whole species and a local population more or less distinctive today may merge next year by interbreeding with its neighbors. The species, however, is a distinct evolutionary unit. It has a role separate from that of any other species. It does not extensively interbreed with other species, and so its variations and its evolutionary changes are not spread through larger taxonomic units. In modern usage it is this fact and not the permitted individual degree of divergence from a type that defines a species.

The basic way in which diversity arises is by splitting of populations, that is, the cessation of interbreeding between populations that did formerly interbreed regularly and that therefore belonged to one species. Thus two species stand where there was one before. The two species now have different evolutionary roles, and changes occurring in one cannot be passed over to the other. In time they diverge, and long continuation of these processes of splitting and of different changes in the separate lines of descent eventually builds up larger groups of more or less similar species. These larger groups are formalized in classification as genera, families, and so on.

Much that has been said about classification up to this point applies equally to fossil and recent organisms. Methods sometimes have to be different in the two cases, but much of paleontology cannot be logically separated from neontology, the study of recent organisms. As regards many principles, it does not matter when an organism lived; it still exemplifies the same basic features of life. There is, however, one fundamental difference. The neontologist studies organisms all living together at a single instant in time. The paleontologist often does this, too, but he also studies populations that succeeded each other through millions of years of time. Both try to decipher the lines of descent among organisms, that is, their phylogeny. The neontologist can do so only by indirect methods. He has no real lines of descent, but only the results of such processes as they appear at a single time. Such indirect methods are also necessary to the paleontologist, but in favorable cases and for various spans of time he has the direct evidence of samples of the successive populations in a real line of descent.

The study of phylogeny is thus particularly dependent on paleontology. Since modern classification is meant to reflect phylogeny and to be consistent with it, classification now also depends largely on paleontology. Not only are there many extinct groups which must be classified, too, but also fossils give the most direct and the only conclusive evidence as to the degrees of relationships of living groups. For instance, hares, rabbits, and pikas, collectively known as lagomorphs, were long believed to be rodents, but the earliest fossil remains of lagomorphs and rodents show that the two groups cannot have had the same origin. In the hierarchy they must be considered different orders. Or, as another example, bears were formerly considered a very distinct group of carnivores perhaps most closely allied to raccoons. Fossils show that bears arose relatively recently (middle to late Miocene) from dogs, to which they are closely related. Their relationship to raccoons is much more distant.

Phylogenetic sequences tend to confirm the evolutionary conclusion that characters-in-common indicate common ancestry, but they lead to great modification of the concept and of its applications. For one thing, they show that characters-in-common often did not really exist in the common ancestors of two descendent groups but were developed *after* the latter groups became separate. The separate rise of such common characters is called "parallelism" or "convergence" and will be mentioned again in Chapter 9. If similar characters have really been inherited from a common ancestor they are called "homologous." If they have arisen separately, they are "homoplastic." If homoplastic characters show similarity of function without homologous structure, they are "analogous."

Birds, bats, and insects all have wings. (Fig. 32.) All bird wings, even to the nonflying wings of ostriches or the flippers of penguins, were derived from ancestral wings like those of *Archaeopteryx,* and so they are all homologous. Bird wings have bones, such as the humerus, the bone nearest the body, derived from bones in the forelimbs of reptiles. Bat wings also contain a humerus, also ultimately derived from the upper bone in the forelimbs of primitive reptiles. The humerus of a bird and that of a bat are homologous. But the bird wing was derived directly from a reptile limb in the Jurassic. The bat wing was derived much later, in the Paleocene, probably, from the front leg of insectivorous mammals. Bird wings and bat wings, as such, are homoplastic and analogous, not homologous, even though they contain homologous bones. The insect wing was of totally different origin. It is analogous to the other two and has nothing homologous to any part of their structure.

Homologous structures are evidence of common ancestry. Homoplastic ones are not. Since homoplastic structures may tend to be more alike in closely than in distantly related forms, they do have some bearing on relationship, but the mere fact that they are not homologous indicates some more or less ancient separation of descent. Much of the process of deciding how nearly related organisms are consists first of seeing how much alike they are and then of deciding how much of the resemblance is homologous and how much is homoplastic. To take a broad and rather simple example, whales

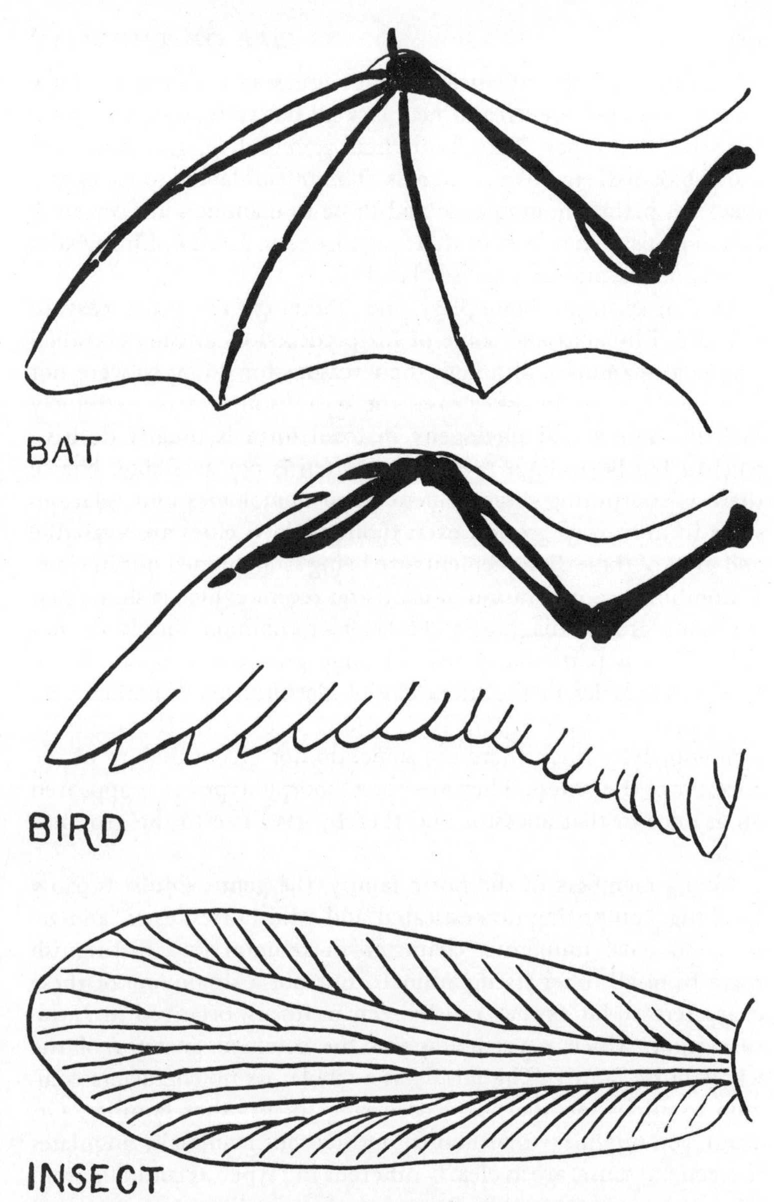

FIG. 32. Wings as an example of homology and analogy. The three types of wings shown are homoplastic and analogous but not homologous to each other. Bat and bird wings contain homologous bones. The insect wing shown is from a Carboniferous forerunner of the cockroaches.

have numerous resemblances to both fishes and mammals. They live in the water, are streamlined, lack all but vestiges of hair, have fins, etc., like fishes. They bear their young alive, give milk, are warm-blooded, etc., like mammals. The resemblances to fishes are, however, plainly homoplastic and those to mammals are certainly homologous. Therefore, in their phylogenetic relationships whales are mammals, and they are so classified.

In that example homoplasy and homology are quite easy to separate. Linnaeus and some of his predecessors already classified whales as mammals, although their reasons for doing so were not the same as ours. In other cases the separation may be extremely difficult. The actual phylogeny in fossil form is usually decisive when it has been discovered, but it often is not available. Hence there is continuing disagreement as to homologies and relationships in numerous groups, even though added clues are available and most of these disagreements are being straightened out in time.

Combined consideration of fossil and recent evidence shows that for many groups diagnostic characters-in-common simply do not exist. This is particularly true of large groups corresponding to higher categories in the hierarchy of classification. Another very important conclusion from such studies is that the characters-in-common, even when there are some, do not give a picture of the ancestor of the group. They are not a "morphotype" that appeared all at once in that ancestor and thereby gave rise to the group as such.

Living members of the horse family, the genus *Equus* broadly speaking, comprising domesticated and wild horses, asses, and zebras, do have numerous characters-in-common that distinguish them from all other living animals. But not a single one of these characters, so far as they can be seen in fossils, occurred in *Hyracotherium,* which nevertheless was the common ancestor of the whole horse family. The family as a whole has no characters common to all its members and occurring in no other families. Or, again, you might say that hoofed herbivorous mammals, ungulates in a general sense, are as clearly different in "type" as could be from the carnivorous mammals. But none of these differences occurred in the earliest ungulates and carnivores, which are exceedingly difficult to tell apart. Yet their actually recorded histories show that the horse family, Equidae, the Order Carnivora, and, more loosely, the various orders of ungulates are valid taxonomic units. "Valid" here means that the way we classify them is consistent with what was almost certainly their real phylogeny as revealed by fossils.

Perhaps what all this adds up to is that evolution really did occur. The various groups of classification did arise on the basis of variation and by continuous change. They were not created in accordance with archetypic patterns.

The fact that it shows succession through long periods of time is the basis for the great contribution of the fossil record to classification and to evolutionary theory. It is, however, also the source of some special difficulties. The most important of these is the problem of successive taxonomic units, as opposed to contemporaneous ones. When a species of animals, for instance, is studied in the recent fauna it is a fixed unit. It does not change appreciably while being studied. In fact it is the rarest of exceptions for any species to have changed in a significant way during the whole period of human observation—evolution is not that rapid. Moreover, in the great majority of cases a recent species is clear cut. It does not regularly interbreed with related species and its variation in some characters does not intergrade with that of any other species.

In the fossil record, on the other hand, it is frequently possible to follow a species through successive strata and in most such cases the species changes quite appreciably. Often the late members differ from the earlier ones as much as one species commonly differs from another in the recent fauna, or differ considerably more than that. The whole series is one *evolutionary* species. It is one, continuing, ancestral-descendent line, a single evolving, self-reproducing population or lineage. At any one time it is also a single species in the sense of neontology. Yet the earlier and later members are different, and different names for them are generally needed by the paleontologist.

The usual practical solution is to call earlier and later parts of the sequence different species and give them different names if they do differ as much as is usual among related contemporaneous species. Then the word "species" is being used in two very different ways. The relationship between an ancestral "species" and a different "species" descended from it is obviously not the same as that between two "species" living at the same time. It has been suggested that paleontologists use some other form for their ancestral-descendent "species," but for various reasons this is not entirely practical and it has not been adopted. Both quite distinct sorts of taxonomic units continue to be called "species," but it is important to remember that they are different. The same sort of distinction exists for genera and other categories.

In cases of such successive species or other units a continuous intergradation through time often exists from earlier to later species. Then a purely arbitrary line must be drawn between the earlier and later species. Each is a real unit in nature, but the division we draw between them did not exist in nature. It is like having a piece of string grading evenly from blue at one end to green at the other. We decide to classify it into a blue piece and a green piece, and so we cut it in the middle. We have separated bluer and greener parts which existed as realities, but the separation is arbitrary, and just where we made the cut one side is the same as the other.

Table 2 gives an example of this sort of situation, one of many examples known to paleontologists. The data refer to small, primitive ungulates (condylarths) from successive faunas in Wyoming (oldest at the bottom in the table). The clearest change is in size, and the table is based on the length in millimeters of the first lower molar tooth. Numbers in the table are numbers of individual teeth in the given size classes in each fauna. The samples are rather small and somewhat irregular in variation, which is true of most fossil samples. Nevertheless they clearly show that in each successive step there was a variable population with a more frequent average size group and rarer variants larger and smaller than this. They also clearly show that in successive faunas the average size, and also the extremes, tended to become rather steadily larger.

It happens that these specimens were classified years ago by the typological method. It was decided that the range of size is such that four species occur and that the "typical" condition for each of the four was to have the length of this tooth about 5½, 6, 7, and 8 millimeters in length, respectively. A type of about this size was designated for each species. The other specimens were then "identified," one by one, by comparing them with the types and labeling each as belonging to the species of the most nearly similar type. For instance, any specimen between about 6½ and 7½ millimeters in length was labeled *Ectocion osbornianus*. The specimens were placed in species as shown by the vertical columns of the table.

That used to be the universal method of classifying specimens and it is still followed by some workers, but this example plainly shows that it is nonsense. All the individuals that lived together in the Clark Fork fauna were parts of the same interbreeding, varying population. To put them in three "species" entirely falsifies the situation in nature. The same is true of the sample from the Gray Bull fauna.

TABLE 2. LENGTH OF M_1 IN ECTOCION FROM WYOMING
(*See text for explanation*)

AGE	FAUNA	SIZE GROUPS								SUBSPECIFIC STAGES IN THE SINGLE EVOLUTIONARY SPECIES
		5.3–5.6	5.7–6.0	6.1–6.4	6.5–6.8	6.9–7.2	7.3–7.6	7.7–8.0	8.1–8.4	*E. osbornianus:*
Early Eocene	Lost Cabin	–	–	–	–	–	–	–	1	*E.o. superstes*
	Gray Bull	–	–	1	7	2	2	1	–	*E.o. osbornianus*
Late Paleocene	Clark Fork	1	1	6	4	1	–	–	–	*E.o. ralstonensis*
Typological "species":		*E. parvus*	*E. ralstonensis*		*E. osbornianus*			*E. superstes*		

In fact the whole sequence is one evolutionary species, or a lineage, changing in time. Among these animals, those living at any one time or in a limited and separable span of time can only be realistically classified as belonging to one species or other taxonomic unit. If these units need to be designated that can be done by naming the whole population for each fauna, as in the horizontal rows of the table. In the example they are named as subspecies because variation of the successive, distinguished populations evidently overlapped widely. This is also a somewhat arbitrary procedure, but it makes sense in terms of what really was going on.

An important point involved here is that interpretation depends not only on the specimens themselves but also on their associations and ages. Meaningful classification of fossils requires information as to their occurrence in the rocks. Without the knowledge that the fossils of Table 2 occurred in specified regions and at specified levels in certain sorts of rocks, there would be no proper way to classify them except as one extremely variable species. That interpretation would still be biologically correct, which the typological classification as four species was not; but it would lose a real and important aspect of the situation: that the species was less variable at any one time and that it changed through time. The same considerations apply to modern classification of living organisms. They are all of the same age, so that variable is fixed, but it is necessary to know where and under what conditions they live.

The modern sample-and-population method also has an important practical advantage that may be mentioned in passing. By typological classification the same species, *Ectocion ralstonensis* and *E. osbornianus,* occur in both the Clark Fork (Paleocene) and Gray Bull (Eocene) faunas. They provide no way to tell the two faunas apart or to correlate rocks in which these species are found. But the *populations* of the two levels are readily separable, with one stage of the evolutionary species occurring only in the Clark Fork and another in the Gray Bull. The ages can thus be distinguished and rocks correlated by means of these fossils.

It was originally planned to end this brief discussion of classification with a summary classification of the more important sorts of organisms, ancient and recent. The summary is, however, rather lengthy: there are and have been so many important sorts of organisms! Since it is a series of thumbnail discussions of one group after another and cannot be given essay or narrative form, some readers may also find it dull. It has therefore been relegated to an appendix at the end of the book. If you are interested turn to it and read it now. If you are not especially interested glance over it, and then use it hereafter as a sort of dictionary when you want to know the general nature of some group of organisms mentioned in later chapters.

8. *Life and Time*

HUNDREDS of thousands of fossils have been collected and studied. The result is surely impressive, but it certainly represents an extremely small fraction of the untold thousands of millions of organisms that have lived since life began. Some students are so impressed by the accomplishment that they insist that the record be taken entirely at its face value, that nothing really essential is lacking. Others are so impressed by the still obvious gaps in the record that they insist that it has little over-all significance and can only be judged as an inadequate series of scattered samples. Both conclusions are unjustified. The record is very incomplete and is misleading if allowance is not made for this fact. On the other hand, such allowance can reasonably be made, and when it is made the over-all record does give reliable information about the general and total evolution of life.

It has been noted (Chapter 2) that many of the nonwoody plants and all of the completely soft-bodied animals which play a large part in the life of today and in its history are rarely and sporadically preserved as fossils. Their fossil records are generally so incomplete as to have little value. Most higher plants and most animals with hard parts have left a fairly good fossil record which has now been quite well sampled by collectors. For most groups the sampling includes a very small fraction of the species that formerly lived on the earth. Any one species tends to be rather local in distribution and different related species tend to occur in different areas. Depending, as it does, on local surface outcrops of ancient rocks, paleontological sampling cannot hope to include all the species of a given group. As an average rule genera tend to occupy larger and more overlapping areas, and this is increasingly true of higher categories, families to phyla. The higher the category the higher the percentage of groups of organisms revealed by paleontological sampling. In many groups the sampling of genera is quite good, often probably including a majority of the genera that ever existed and clearly giving an adequate idea of the general diversity of each group.

When followed in detail over long periods of time the precise record of any one group usually reveals various gaps or apparent jumps. There is some dispute about this feature of the record, but the consensus is that it is usually or always due to two causes. First, continuous deposition of sediments (later available for study at the surface) over very long periods of time is the exception and not the rule (Chapter 3). Interruption in sedimentation, represented by an unconformity in the rocks, also means a gap in the fossil record. Second, evolving groups of organisms do not stay in the same places or even necessarily in the same quarters of the earth or the same broad sorts of environments throughout their histories. In a rock sequence of one facies or in one region movements of groups of organisms into or out of the environmental facies or the geographic area appear as abrupt changes in the record. Collection in many different facies and regions helps greatly to complete such records, but nowhere near all facies and all areas at all times are represented by rocks now accessible for collecting.

Certain sorts of environments and of organisms, even among those with hard parts, are consistently poorly represented in the record. In mountains and uplands erosion usually predominates over sedimentation. Aside from strays or scraps washed into the lowlands the inhabitants of such regions are rarely permanently preserved in sedimentary rocks. In the abysses of the oceans slow sedimentation goes on all the time and erosion is slight, but deposits of the abysses have only very rarely—some would say never—been uplifted to where productive methods of fossil hunting are possible. Among inhabitants of environments otherwise well represented by fossils, some are not well sampled by the processes of death and burial. Small flying animals, such as bats, most birds, and most insects, are usually less well represented among fossils than are larger and ground-living animals. This is generally true in what might be called normal conditions of fossilization, even though otherwise uncommon fossils may be abundant in sporadic and special circumstances, such as cave deposits for bats or amber for insects.

All these defects are serious and their result is that the fossil record cannot be taken at its face value as a total picture of the history of life. Yet attention to these defects and to the many sorts of clues available as to their existence and nature has taught us how to make allowances for them in large part. Within limits, becoming narrower as knowledge increases, it is possible to judge how good our samples of past life are, that is, what proportion of the sorts of living organisms are actually known by fossils. It is also possible with increasing accuracy to infer what occurred during gaps in the record. Still another defect, the fact that almost all fossils are parts and not whole organisms, can also be in large part overcome by improving methods of inference (Chapter 4).

At present the most serious and glaring defect of the record as a whole is our extremely inadequate knowledge of pre-Cambrian life. From early Cambrian to Recent, a respectable span of some 500 million years, the fossil record with all its defects is rich and essentially continuous. We know, however, that life must have existed for at least twice and perhaps three times that long before the Cambrian. For all that tremendous early time in the history of life the fossil record is so poor as to be almost negligible. Discovery of many different kinds of pre-Cambrian life has been claimed from time to time. In most cases, however, there are serious doubts either as to the age or as to the identification of supposedly pre-Cambrian fossils. The only ones that now seem reasonably certain are calcareous algae, known as far back as the middle pre-Cambrian or somewhat earlier. Life was therefore in existence at that very remote time, at least 1000 and perhaps as much as 1500 million years ago. Algae, although very primitive in terms of the total development of life, represent an enormous advance over what the earliest sorts of life must have been like. The origin of life presumably occurred long before algae appear in the fossil record.

For plants the occurrence of pre-Cambrian algae may correctly indicate the level of evolution then reached. The other, higher major groups (phyla) of true plants, all of which are primarily nonaquatic, probably were post-Cambrian in origin. All of them appear in the fossil record in the Silurian or Devonian, and there is no reason to infer that their real origin was much earlier. (One exception is the Phylum Bryophyta, liverworts and mosses; they have a poor fossil record and might be older than the Silurian, but they are not known before the Carboniferous and certainly need not be assumed to have arisen in the pre-Cambrian.) The general record of plants is summarized in the appendix of this book.

No pre-Cambrian animals are *surely* known, although a few claimed occurrences may eventually be substantiated. Yet the early

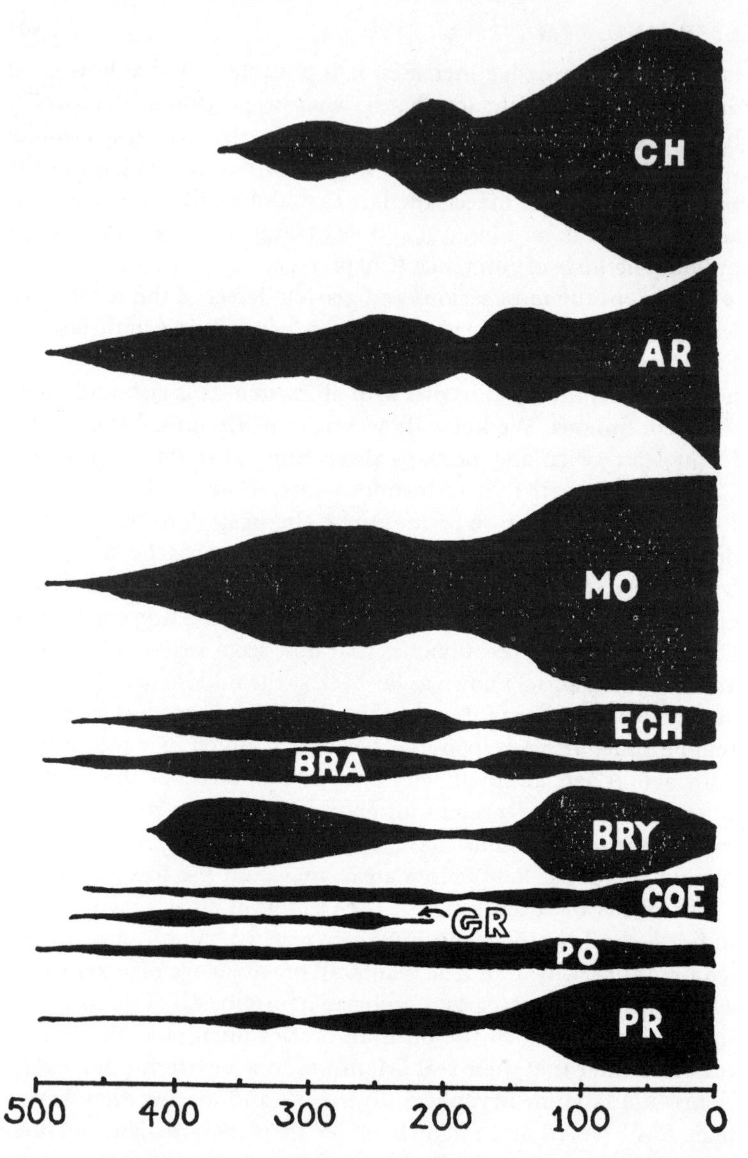

FIG. 33. The fossil record. Relative known variety of various phyla of animals. The time scale at the bottom indicates millions of years before the present. AR, arthropods: crabs, spiders, insects, and their allies. BRA, brachiopods: lamp shells. BRY, bryozoans: moss animals. CH, chordates: fishes, amphibians, reptiles, birds, and mammals. COE,

Cambrian faunas include, among other animals, arthropods, which are the most advanced and complex of invertebrate animals, even though the Cambrian forms are decidedly primitive in comparison with later arthropods. Certainly animals must have been evolving for a long time in the pre-Cambrian. All the phyla of animals—so far as the groups clearly deserve this rank and are fossilizable at all—had appeared by the end of the Ordovician. It does not follow, as often claimed, that all were pre-Cambrian in origin. Cambrian and Ordovician were long periods, spanning surely 100 million and perhaps as much as 150 million years between them. During that time the fauna was progressively enriched in major groups, at least some of which were probably just then arising, and it is not true that all or most of the animal phyla suddenly appear just at the beginning of the Cambrian. Nevertheless the absence of pre-Cambrian animal fossils and of fossils revealing basic stages in origins of the animal phyla is a serious problem. The most probable explanation is that organisms in the early stages were all individually very small and soft-bodied animals.

The general nature of the fossil record of animals from Cambrian to Recent is shown in Fig. 33. The diagram approximately indicates the *known* variety of groups, especially genera, in each phylum through the successive geological periods. The phyla are characterized, their main subdivisions mentioned, and some other details of their fossil records summarized in the appendix. For reasons suggested earlier in the chapter it cannot be supposed that this diagram of what is known is an entirely accurate representation in absolute terms of what really occurred in nature. For instance, the wormlike phyla were certainly more abundant throughout than the factual record shows, and the Arthropoda surely expanded much more in the later half of their history than is indicated. Nevertheless the relative variety for most of the phyla is probably correctly indicated, and the expansions and contractions within each phylum represent real trends in most or all cases.

---

coelenterates: corals and their allies. ECH, echinoderms: starfish, sea urchins, sea lilies, and their allies. GR, graptolites: an extinct group of uncertain affinities. MO, mollusks: snails, clams, and their allies. PO, poriferans: sponges. PR, protozoans: noncellular animal-like organisms. Further description of the various groups is given in the appendix.

It is interesting that no phylum has expanded steadily from the time of its appearance to the present day. All show from two to four different periods of expansion, separated by periods when the phylum was somewhat (for example, Mollusca in the Permian) or very much (Bryozoa in the Triassic) less varied. Apart from this, the patterns of expansion and contraction are diverse and quite different for each phylum. The most nearly general feature is that most of the phyla contracted in the Permian, Triassic, or both.

If the whole record is added together it appears that the diversity of animal life increased greatly from early Cambrian into the Silurian, fluctuated without really marked changes from Silurian to Permian, declined in the Triassic, and then expanded greatly again from Jurassic into Tertiary. It may now have contracted slightly again but still is much greater than at any time in the history of life before the Cretaceous, at least. There has been a general but clearly not a constant tendency for life to become more varied, that is, to include more different *sorts* of organisms, since the early Cambrian.

Whether this means that life has also become more *abundant,* whether the total mass of living matter has increased, is less apparent. No way of estimating this total mass from the fossil record, even roughly, has yet been worked out. Increase in diversity does not necessarily correlate well with increase in mass. Among ecologically similar organisms the mass is about the same if many individuals of one species or few individuals of each of many species play a given ecological role. Increase in diversity implies some increase in the number of ecological roles and so would be expected to lead to some increase of mass in spite of the fact last mentioned. However, the total amount of energy and material in the system of life is almost completely limited by the amount of photosynthesis performed by green plants. This, in turn, is limited by the amount of solar radiation and the efficiency of photosynthesis. It is highly improbable that either of these factors has increased appreciably since the Cambrian, at least, and probably long before.

It thus seems improbable that any really noteworthy increase in the mass of life can occur in a given habitat once that habitat is fully occupied by even a few species of plants and animals. The aquatic, or at least the marine, habitats may have been almost fully occupied by the end of the pre-Cambrian and seem rather surely to have been by the end of the Ordovician. Occupation of land habitats, however, did not begin until the Silurian, as far as the fossil record shows. There may have been terrestrial microorganisms much earlier, but if so it seems unlikely that larger plants and animals would have been long delayed in following them onto the land. In the Carboniferous lowland habitats were well occupied but uplands probably were not. Essentially full occupation of the land was apparently completed during the Mesozoic, perhaps sometime between Jurassic and late Cretaceous. The process of occupation of the land must greatly have increased the mass and not only the variety of life. Maximal increase in variety seems to follow attainment of maximum mass; both are apparently now near their highest points for the whole history of life. Whether further increase is possible or will occur is a matter of opinion without much sound basis for prediction.

(One pertinent factor not noted in the preceding discussion is rate of turnover of materials in living communities. The actual mass present at any one time will be lower if turnover is faster. Although some animals now may have higher metabolic rates, or turnover, than any in the Cambrian, I see no good way of judging whether average turnover for the whole earth has changed appreciably.)

When the records for particular groups are more closely examined it is seen that most of them have fairly well-defined episodes during which they were diversifying with unusual rapidity. These episodes may come at any time in the history of the groups. There is apparently some tendency for them to occur, or at least to culminate, during the first half but not at the beginning of that history. It is not uncommon for a broad group of animals, such as an order or a class, to have two, three, or even more such episodes. The phenomenon is generally called "explosive evolution," but the term is possibly misleading for the whole of such an episode seldom occurs in less than 10 and may take 50 or more million years—a slow explosion, indeed, by human standards!

The phenomenon is more complex than is suggested by such curves, but one aspect of it is readily seen by plotting the numbers of new genera appearing in the fossil record per million years. Such a plot is shown for the whole Phylum Brachiopoda in Fig. 34. Sustained high rates are shown from Ordovician to early Carboniferous, and then there is another strongly marked climax of explosive evolution in the Jurassic. These episodes can be related to the total diversity of the group at various times, as shown in Fig. 33. Further analysis shows that the sustained (for well over 100 million years) explosion in the Paleozoic is really the sum of a whole series of lesser explosions within various different groups of brachiopods.

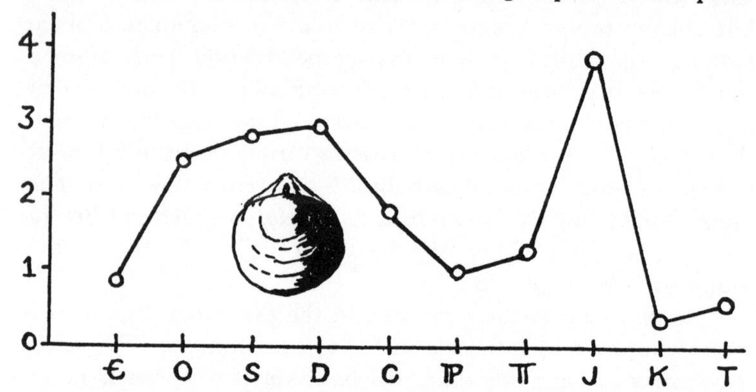

FIG. 34. Rates of evolution (diversification) of brachiopods. The scale to the left indicates the known number of new genera arising per million years. The symbols at the bottom indicate the geological periods in sequence from Cambrian to Tertiary. The horizontal distances are not proportional to absolute time.

Fig. 35 is a similar graph for seven of the eight classes of vertebrates. It shows that the development of the phylum as a whole, including successively greater expansions in the Devonian, Permian, and Tertiary (Fig. 33), is in large part a resultant of a complex series of explosions in its various classes.

The relatively rapid diversification involved in an explosion begins when for any reason an evolving group of organisms has access to a new broad sort of environment or acquires a new and particularly useful way of life: the two often go together. The Permian expansion of vertebrates involved mainly their spread to land environments after the rise of the first fully terrestrial vertebrates, the reptiles. The Tertiary expansion was mainly due to the physiologically superior adaptations of a different group of fully terrestrial vertebrates, the mammals. The example shows that explosion need not soon follow origin of the new adaptation or way of life, for mammals had existed for at least 75 million years before their most clearly explosive expansion began.

Another major feature of the fossil record, and correspondingly of the history of life, is that of succession and replacement of one group of organisms by another. This is seldom evident in the history of the animal phyla (Fig. 33). The reason is that these broad

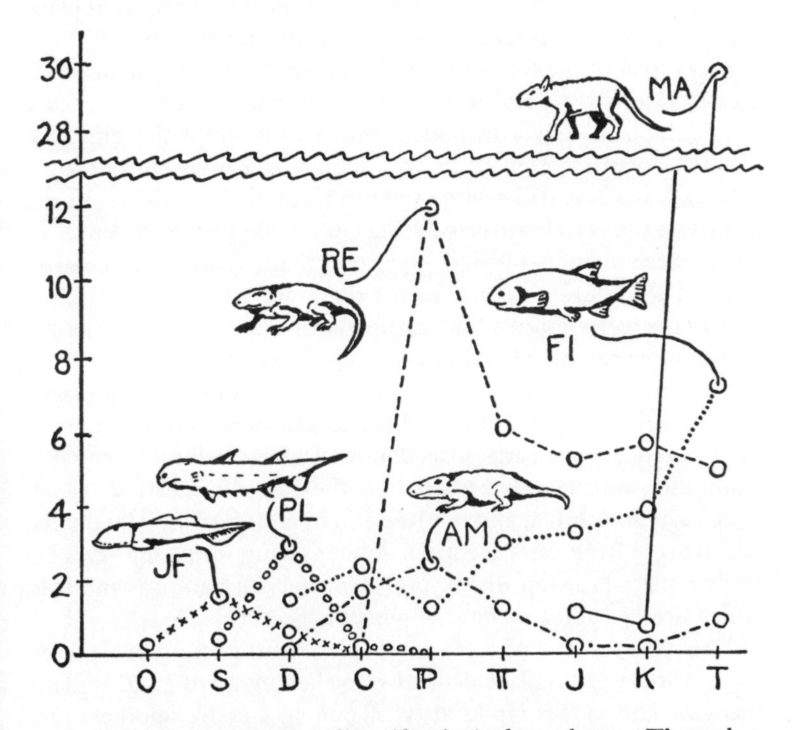

FIG. 35. Rates of evolution (diversification) of vertebrates. The scale to the left indicates the known number of genera arising per million years. The symbols at the bottom indicate the geological periods in sequence from Ordovician to Tertiary. (The horizontal distances are not proportional to absolute time.) AM, amphibians. FI, true fishes (Chondrichthyes and Osteichthyes together). JF, jawless "fishes" (Agnatha). MA, mammals. PL, placoderms, an extinct group of prefishes. RE, reptiles. Each symbol is connected to the point of most rapid diversification of the corresponding group. (Birds diversified most rapidly in the Tertiary; their fossil record is too poor to give meaningful figures in such a diagram.)

groups represent basically different ways of life. Occasionally members of one phylum may become ecologically similar to those of another, but on the whole their roles and potentialities are so different that replacement of one by the other is out of the question. Some mollusks and some brachiopods are enough alike so that there is a certain relationship between expansion of one phylum and contraction of the other, but mollusks could not possibly replace the Bryozoa, for instance. (Among plants the pteropsids have largely replaced the lycopsids and sphenopsids, but the plant phyla are not closely comparable to those of animals; in fact, although I have designated phyla in both kingdoms for simplicity and uniformity, botanists often do not call their major divisions phyla.)

On a lesser but still a large scale, replacement is evident in the history of the vertebrate classes (Fig. 36). Aside from differences in detail three major ecological types occur. In terms of locomotion adapted to different environments there are swimming, walking, and flying vertebrates. The earliest swimming forms, Agnatha, were almost completely replaced by the Placodermi, derived from early Agnatha. The Placodermi in turn were completely replaced by Chondrichthyes and Osteichthyes, together and not in sequence, both derived from early placoderms. Among walking forms the Amphibia were largely replaced by the Reptilia, derived, again, from early Amphibia, and the Reptilia in large part by Mammalia, which arose from early Reptilia. Among flying forms the Aves entirely replaced the reptilian flyers (pterodactyls), but birds and bats, which are of course mammals, persist together.

The phenomenon of ecological replacement goes on at all levels from world-wide replacement of one class by another to replacement at one particular locality of one species by another. One other example, intermediate as to geographic extent and as to the scope of the groups involved, is given in Fig. 37. The proportions of various groups of artiodactyls, even-toed hoofed mammals, in the North American fauna have changed constantly as one group replaced another. Of the four groups present in the Eocene not one survives in North America today, although two of them do survive elsewhere. As would be expected, the three groups surviving in North America have quite different ecological roles. The bovoids (bison, mountain "goats" and sheep, pronghorn "antelopes") are grazers of open plains and mountains; the cervoids (deer) are browsers of the forests; and the suines (peccaries in North America) are piglike omnivores now confined to warm lowlands.

This sort of parceling out of ecological roles explains why a later large group does not always completely replace a similar earlier group. In the example given earlier the reptiles replaced most

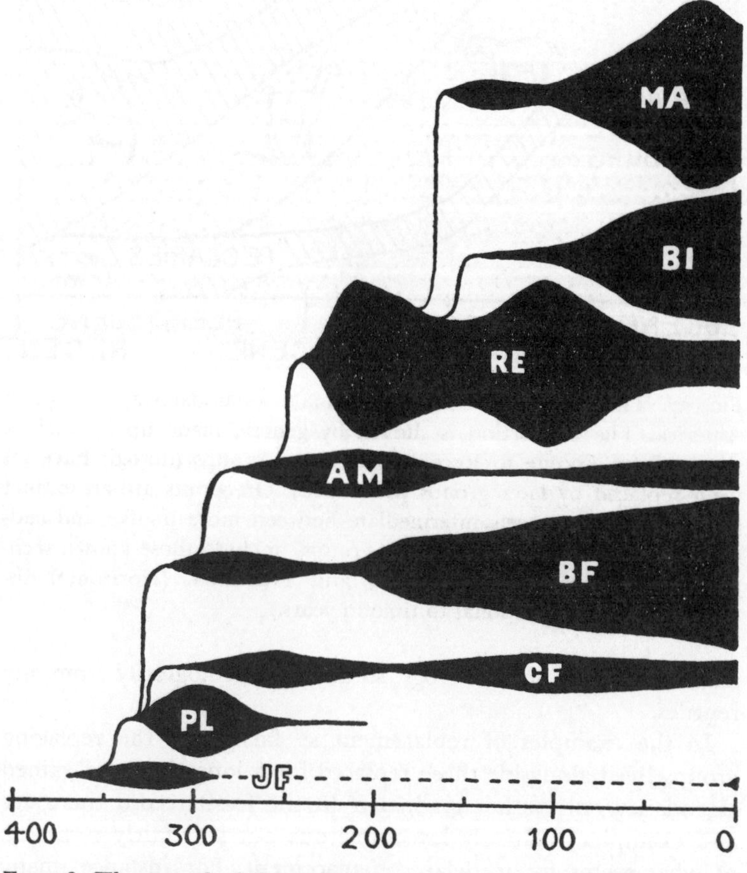

FIG. 36. The record of the vertebrates. Relative known variety of the classes of vertebrates. The scale at the bottom indicates millions of years before the present. AM, amphibians. BF, bony fishes (Osteichthyes). BI, birds (Aves). CF, cartilage fishes (Chondrichthyes). JF, jawless "fishes" (Agnatha). MA, mammals. PL, placoderms, an extinct group of prefishes. RE, reptiles.

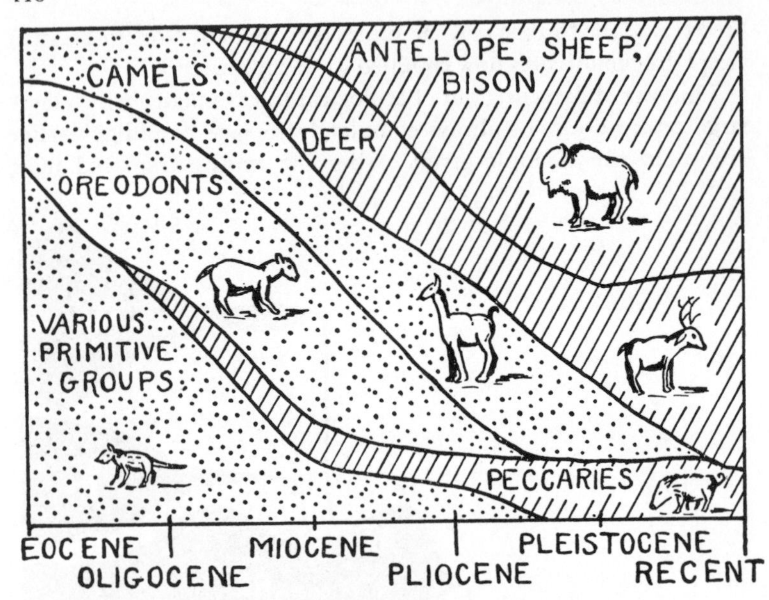

FIG. 37. The even-toed hoofed mammals (Artiodactyla) in North America. The proportion is shown, by genera, made up by various groups from Eocene to Recent. The oldest groups (dotted) have all been replaced by later groups (crosslined). Oreodonts are an extinct group, in some respects intermediate between more piglike and cud-chewing types. "Various primitive groups" include those known technically as Palaeodonta, Ancodonta, and Tragulina. (Horizontal distances are not proportional to time in years.)

amphibians but not the frogs, so different ecologically from any reptiles.

In the examples of replacement so far given, the replacing groups lived alongside those replaced for a long time and gained ascendancy only rather gradually. In the fossil record there are also examples, relatively less numerous but still fairly common, of what seems to be delayed replacement. For instance, many cetaceans (whales and their allies, mammals of course) are so like ichthyosaurs (marine reptiles) that they must have similar ecological roles. Yet the ichthyosaurs became extinct some tens of millions of years before similar cetaceans appeared, and in the meantime there were no animals at all like these in the seas. In still other cases the expansion of one group follows the extinction of another so promptly, as geological time goes, that one cannot but think of cause and effect; and yet the groups do not seem to have been in direct competition. Replacement of reptiles (many of them, at least) by mammals was really an example of this sort. Mammals and dinosaurs lived together for a long time but were then so different that any significant competition is almost unthinkable. Yet the major expansion of mammals began immediately when dinosaurs became extinct, and *later* there evolved mammals that would have competed with dinosaurs if there had been any dinosaurs left alive.

Looking back over the whole record we see that life began at some unknown and almost unimaginably remote time in the early pre-Cambrian. The earliest well-organized forms of life were surely protistans, noncellular forms of "animals" and "plants" that may better be considered a separate kingdom (see the appendix). From them arose aquatic true plants, primary producers, and animals, consumers of plants and of each other. The primarily aquatic plants never developed much fundamental diversity. They came to include many species and genera, to be sure, but their fundamental structural plan, such as characterizes a phylum of animals, was simple and not much varied. That plan was already well-established long before the Cambrian and has never changed much. In the Silurian and Devonian true, multicellular plants invaded the land and then underwent a basic diversification that did produce several fundamentally different physiological and structural groups. But this *basic* diversity never became as great as in animals.

In marked contrast with plants all the really basic differentiation of true, multicellular animals took place in aquatic forms. Although not very rapid in human terms that differentiation apparently all occurred during a rather limited part of geological time, perhaps 200 million years, toward the end of the pre-Cambrian and during the Cambrian and Ordovician. Invasion of the land, beginning in the Silurian for invertebrates and the Devonian for vertebrates, eventuated in a great number of new lesser groups, literally millions of them, but no really basic anatomical changes.

In life in general, protistans, plants, and animals alike, there has been an irregular and yet evident general tendency for increase in diversity. In this respect life is near its maximum today. One reason for this tendency has been a spread to quite new habitats.

Most notable on the broadest scale was the spread from water to land. Within generally occupied habitats there has also been a tendency for each sort of organism to become more local in distribution, more specifically adapted to a particular part of the habitat and to a special ecological role in it. This tendency, again, has been quite irregular and subject to numerous exceptions, but it does seem to have been present in the average, long-term, over-all evolution of life.

Within the various major groups of life there has also been a variable and yet quite widespread tendency for replacement to occur, for older groups to cede their ecological roles to newer ones. Often the replacing forms originated from early members of the group eventually replaced. Often the replacement was by direct competition. Sometimes it involved no apparent competition and it could be long delayed. Replacement of species, sooner or later, has been the rule. Among higher and higher categories replacement has been less and less common. It has not occurred among the highest animal categories, the phyla.

Spread into new habitats, increasing diversification, progressive changes of evolving lineages, and replacements of one group by another are the great movements in the history of life. The panorama is tremendous. The long sequence has continuously and profoundly changed the whole aspect of life on earth. With the coming of man an essentially new element was introduced. The old processes do not cease to operate, but with man present the evolution of life has become quite different in some respects. Discussion of this point, so important to us, is deferred to Chapter 12.

9. *Ways of Organic Change*

ONLY AMONG living organisms is it possible to study in detail the basis for their variations and the ways in which these are passed on through the generations. Results of these processes are evident in fossil populations. Fossils always vary, and they vary in about the same degrees and ways as recent populations. Here again the present is a key to the past. It is clear that the origin and inheritance of variation have had the same nature throughout the history of life. Discovery of the details of these processes has been the greatest contribution of neontology to an understanding of evolution.

On the other hand, it is a severe handicap not to be able to follow recent populations through spans of time long enough for significant evolutionary change to occur in them. Man has been studying animals and plants in a rather haphazard way for a few thousand years and in a really systematic way for about four hundred years. In a few exceptional cases small changes have been observed in species in nature, things like darkening of the average color or increased resistance to insecticides. Such changes can be hurried up somewhat by artificial means. The whole art of plant and animal breeding consists of hastening evolution and directing it, as far as nature presents materials, in directions desirable to man. But, to take an extreme example, a dog is still a dog and of the species *Canis familiaris* whether it is a Pekingese or a great Dane. We now know that in nature change from one species to another usually takes some 500 thousand years and often much longer. It is impossible for man directly to observe anything so slow in terms of human generations, and equally impossible to hasten it experimentally to any sufficient degree without losing touch with the realities of the situation in nature.

The greatest contribution of paleontology to the understanding of evolution is that it permits us to see what we cannot see in living organisms: the actual changes in populations under completely natural conditions and over periods of millions of years. Without this we never could know whether the factors found in the laboratory really are significant for the history of life or how, in fact, evolution really does occur.

A first essential point demonstrated by the fossil record is that change normally or at least usually occurs gradually in varying populations. Every population has considerable variation in many respects. As successive, ancestral and descendent populations change in time, one sort or direction of variation becomes progressively more common in the population. In populations separated by shorter spans of time variation overlaps more or less widely. For instance, if size is increasing, the descendent population still includes individuals of the same size as most of its ancestors, even though the usual or average size is larger; some individuals are larger than any of the ancestors, and none is quite as small as the smallest ancestors. If the change continues eventually overlap ceases: the smallest descendant is larger than the largest ancestor.

Change of this sort has already been seen in Chapter 7 (Table 2). Another, even better example is differently presented in Fig. 38, which shows a rather radical change of shape in Jurassic oysters. The important point is that in the thousands of known examples for which we have closely successive, surely ancestral and descendent fossil populations change *always* occurs in this way. This is true not only for such characters as size (Table 2) or shape (Fig. 38) but also for the rise of quite new characters and structures. For instance, in the evolution of the horse one of the important changes was the development of an entirely new tissue, cement, on the teeth. This appeared first as the merest film on the teeth of a few individuals in populations otherwise without cement. It became progressively more usual in successive populations and also, on an average, thicker. Finally it occurred and was thick in all members of later populations. Even when the change itself was rather abrupt its evolutionary spread in populations was not. In the horses, again, functional change from three-toed to one-toed feet was relatively rapid and not, as usually represented, a long-continued trend toward reduction of the side toes. Nevertheless, functional side toes were not lost overnight but in a sequence of overlapping populations in which some had working side toes and some did not. The latter condition gradually came to characterize whole populations.

There is still some argument about this point. The view I have presented is that of most students nowadays, and no one can deny the plain fact that gradual change through successive, varying populations is extremely common and is represented by innumerable,

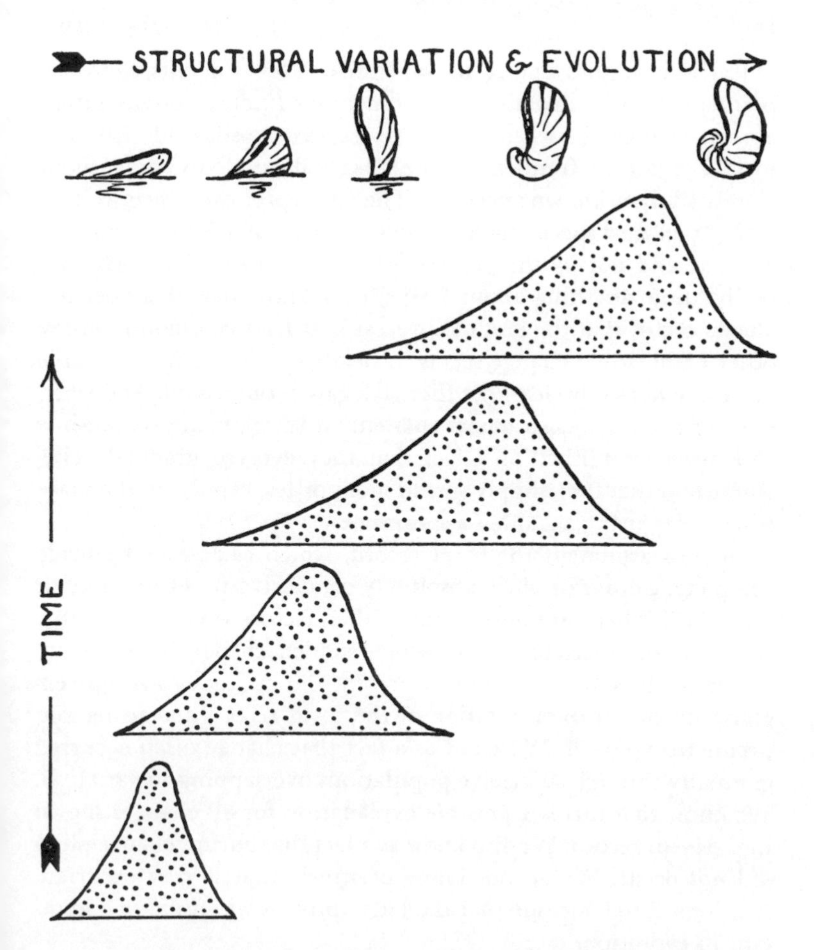

FIG. 38. Evolution of populations of coiled oysters (early Jurassic). The sketches at the top show progressive stages in coiling. The first three are attached, the last two free (see Fig. 11). Any degree of coiling between these stages can occur; they only exemplify a continuous scale. The curves below represent four populations, successive from below upward. The height of a curve at any given point is proportional to the number of individuals with a degree of coiling exemplified vertically above. The four populations shown are also merely representative stages of what was a continuously changing sequence in time. (Data essentially from Trueman.)

objective examples among known fossils. A few students, however, point to the fact that there are sometimes (indeed, often) rather abrupt changes between successive, related fossils and that new sorts of organisms frequently appear as fossils suddenly and without closely similar, known ancestors. They interpret these facts as indicating that new species, genera, and so on often arise suddenly, all at once, and not by the gradual process outlined above. (If they really know anything about fossils, they know that this does not *always* occur, that the gradual process is at least common.) On the other hand, when change seems to be abrupt it is always possible and often perfectly clear that there is a gap in the record. And when new sorts of organisms appear suddenly it is, again, always possible and often established as a fact that they evolved gradually elsewhere and that they simply spread more or less rapidly to the place where they make a sudden appearance.

On this argument the fossil record, which can never be made complete, cannot produce absolutely conclusive proof one way or the other. There are also various other factors involved, notably the supposed mechanisms by which sudden changes occur in individuals. It is, however, fair to present the matter as a simple exercise in logical interpretation of the fossil record and to let you decide for yourself. We know as a fact that change *often* occurred gradually through successive populations overlapping in variation. We know that this is a *possible* explanation for all changes shown in the fossil record. We also know as a fact that abrupt change often did *not* occur. We do not know positively that it *ever* occurred. Is it logical to conclude that the latter process was usual or important in evolution?

(The reader will have noted that this argument is related to the philosophies of classification discussed in the last chapter. Is evolution a matter of sudden appearance of successive "types" or is it a process of change in populations?)

Change of the same sort often continues in the same group of organisms for long periods of time. For instance, a group of animals may become rather steadily larger in average individual size over spans in the tens of millions of years. Such long-continued change in one direction is called a trend and trends are very common in the fossil record. Some paleontologists and many nonpaleontologists have seriously misinterpreted this fact. The correct interpretation is now firmly established and has long been well known to most paleontologists, but it has not yet penetrated into all the general works on life or on evolution. The old, certainly incorrect interpretation must therefore still be mentioned here. That interpretation is that trends are universal in evolution, that once started a trend keeps right on without deviation until the group becomes extinct, and that trends are caused by some mysterious inner force usually called orthogenesis.

The word "orthogenesis" has been used in so many different ways that it really has no exact meaning any more and is being dropped from the vocabularies of careful paleontologists. In the meaning given it in the last paragraph, which is as common usage as any, it is now conclusively known that there is no such thing as orthogenesis. As you really see them in the fossil record, trends do not at all conform to that idea. Just to clear up this old misunderstanding two of the classic supposed examples of orthogenesis may be briefly considered: sabertooths and horses. The sabertooths, which were not really tigers but a large, sharply distinctive group of catlike carnivores, are especially characterized by their great, rather saber-like, canine teeth. Even today, writers not familiar with the actual fossils continue to say that the canines became steadily, "orthogenetically" larger through the history of the group and finally became so overgrown that the sabertooths became extinct. As a matter of plain fact that is completely untrue. Students of the sabertooths have been pointing this out in vain for at least forty years: another example of the durability of error. The earliest sabertooths had canines relatively about as large as the last survivors of the group. For some forty million years of great success the canines simply varied in size, partly at random and partly in accordance with individual advantage to species of various sizes and detailed habits. The famous trend for the sabers to become larger did not really occur at all.

The evolution of the horse family included, indeed, certain trends, but none of these was undeviating or orthogenetic. The uniform, continuous transformation of *Hyracotherium* into *Equus*, so dear to the hearts of generations of textbook writers, never happened in nature. Increase in size, for instance, did not occur at all during the first third of the whole history of the family. Then it occurred quite irregularly, at different rates and to different degrees in a number of different lines of descent. Even after a trend toward larger size had started it was reversed in several groups of horses which became smaller instead of larger. As already briefly noted, the famous "gradual reduction of the side toes" also is something that never happened. There was no reduction for the first 15 or 20 million years of the history. Then there was relatively

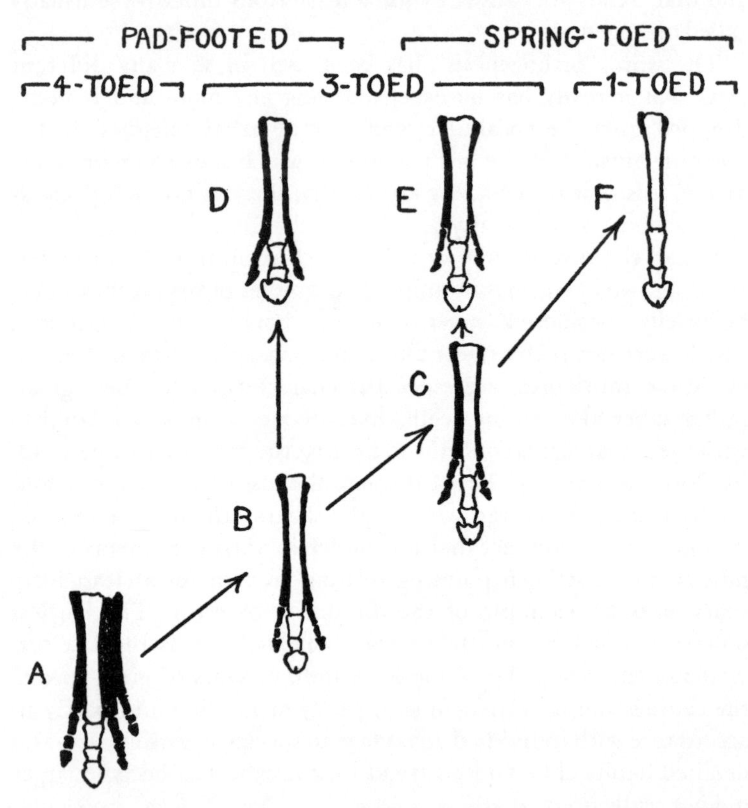

FIG. 39. Evolution of the four sorts of front feet in the horse family. A, *Hyracotherium* (eohippus), early Eocene. B, *Mesohippus,* middle Oligocene. C, *Merychippus,* middle Miocene. D, *Hypohippus,* early Pliocene. E, *Hipparion,* early Pliocene. F, *Pliohippus,* early Pliocene (the foot of the living horse, *Equus,* is essentially similar). There are many other lines of descent in the family and many other horses with each of these sorts of feet.

rapid reduction from four front toes to three (the hind foot already had only three toes). Many horses simply retained the new sort of foot without further change. In one group there was later another relatively rapid change of foot mechanism involving some reduction in size of side toes, which, however, remained functional. Thereafter most horses retained this type of foot without essential change. In just one group, again, another relatively rapid change eliminated functional side toes, after which their descendants simply retained the new sort of foot. (Fig. 39.)

In the history of the horse family there is no known trend that affected the whole family. Moreover, in any one of the numerous different lines of descent there is no known trend that continued uniformly in the same direction and at the same rate throughout. Trends do not really have to act that way: they are not really orthogenetic.

(The evolution of the horse family, Equidae, is now no better known than that of numerous other groups of organisms, but it is still a classic example of evolution in action, and a very instructive example when correctly presented. I do not propose to go into it more fully here. A recent and, I hope, easily understood work on the subject is cited at the end of this book.)

It happens not infrequently that different evolving populations or lineages independently go through similar trends. This is especially likely and the trends may be particularly similar if the lineages in question are more or less closely related. Related lineages splitting off from a common ancestry at about the same time may show similar and simultaneous trends. This phenomenon is that of "parallel trends" or "programme evolution." In other cases the lineages may arise successively and have similar trends one after the other: "successive trends" or "iteration." The two cases intergrade and in principle the phenomenon is the same in both. An example of parallel trends is shown in two groups of brachiopods (see appendix) in Fig. 40. These two groups—they are classified as families—are rather closely related and similar in most respects except for the remarkable difference that their shells curve in opposite directions. The history of each family is rather complicated, but in both families there was a trend for the hinge line, along which the two shells are articulated, to develop numerous small serrations. These appear first toward the middle of the hinge and finally occupy the whole of it. With some irregularity the trend went on for about 100 million years, which is unusually long for any single trend.

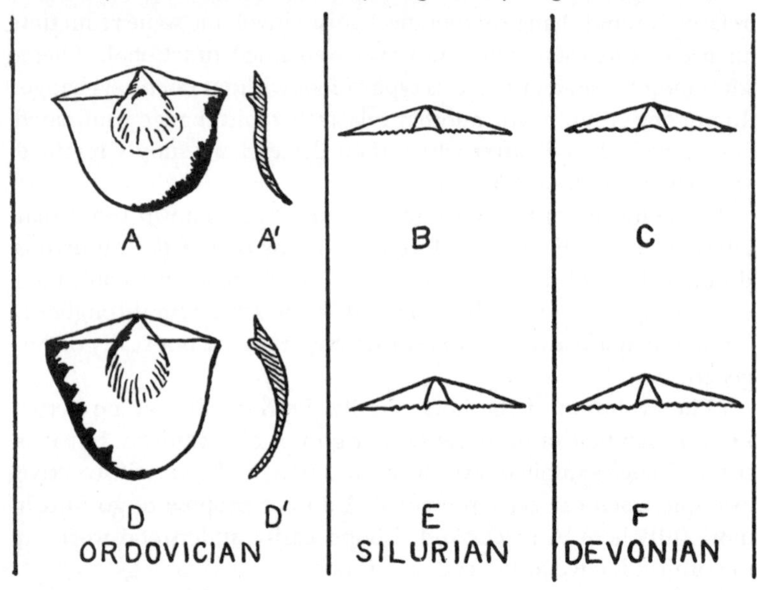

FIG. 40. An example of parallel trends. Each horizontal row represents a different family of brachiopods (upper, Strophomenidae; lower, Stropheodontidae). A and D are internal views of one of the two shells. A′ and D′ are vertical median sections through both shells, showing that the curve is in opposite directions in the two families. B–C and E–F are partial views, corresponding with the upper part of A and D. The hinge line (the horizontal line on each sketch) is seen to be plain in A and D and to become progressively serrated in each family independently, B–C and E–F. The generic names are of technical interest only but are given for reference: A, *Strophomena;* B, *Strophonella;* C, *Strophonelloides;* D, *Rafinesquina;* E, *Brachyprion;* F, *Stropheodonta.* (Essentially after Cloud, but considerably modified.)

In this and in almost all examples of parallel or iterative trends three sorts of characters are distinguishable. There are, first, characters that distinguish the two or more different groups, characters involved in their origin and separation from each other. Such characters usually tend to be retained in each group without much subsequent change. Second, there are characters that arise similarly but separately in the two groups. Their rise is, of course, the feature of parallel or iterative trends. Third, as time goes on each group usually develops other characters, which may also be involved in trends, that are different in the two groups.

Parallel trends are a special case of parallel evolution in general. It is a frequent but not a universal phenomenon of the fossil record that similar organisms tend to change in similar ways. For instance, in the rodents, the most exuberantly diverse order of mammals, many different lineages separately developed more or less similarly complicated patterns of enamel loops and crosscrests on the grinding teeth. Those teeth also tended to become higher or, eventually, to grow continuously without forming roots. (Some rodents, however, changed in quite different ways, and those different changes also commonly occurred in parallel in several different lineages.)

Definition of parallel evolution as similar change in similar organisms implies that the groups concerned were related to begin with. Essential similarity among organisms is a suggestion, at least, of fairly close relationship. The general idea of parallel evolution is that the different lineages changed and yet changed in such a way as to remain similar, without becoming notably more or less so. It also frequently happens that quite dissimilar groups change in similar ways and come to resemble each other more and more closely. This is the evolutionary phenomenon of convergence. All organisms are similar in some respects and dissimilar in others; the terms are relative, and all organisms are related in some degree, close or remote. There is no absolute distinction between parallelism and convergence, but in the most clear-cut cases it can be seen that convergence increases similarity and parallelism does not.

Brachiopods, pelecypods, and corals belong to different phyla (Brachiopoda, Mollusca, Coelenterata) and for the most part they are so obviously distinct that no one could possibly mistake one for another. Yet in each group some (extinct) forms developed the form of a cone, attached at the inverted apex. These forms from the three different phyla, about as distantly related as animals can be, are so similar in external appearance as easily to be mistaken for each other and to require study of the internal anatomy for their distinction. Among many other striking examples of convergence is that of the South American marsupial *Thylacosmilus* and the sabertooths. Their origin was very different—they belong to different infraclasses—but the canine is saber-like in both and there are remarkable similarities in the whole build of the head. (The body is unknown in *Thylacosmilus.*) The resemblance between ichthyosaurs and some cetaceans is a famous case of convergence, previously mentioned in another connection. The resemblance extended even to the mode of birth of the young (Fig. 41).

FIG. 41. Convergence. Above, an ichthyosaur, extinct marine reptile; below, a porpoise, living marine mammal. Both animals are females shown in the act of giving birth to young. In both the young are being born tail first (but young of the dolphin group are sometimes born head first). Although ichthyosaurs were reptiles they bore living young, as do recent sea snakes, as an adaptation to marine life. Fossils have been found of ichthyosaurs in the act of giving birth; these probably represent miscarriages in the death struggle of the mother, but the manner of birth is doubtless the same as in normal cases. Note the strong convergence in form between the two animals, but note also differences: tail fin vertical in ichthyosaurs, horizontal in porpoises; hind flippers in ichthyosaurs, none in porpoises.

Convergence is the bane of the taxonomist. It involves resemblances, sometimes extensive and detailed resemblances, that are not evidence of relationship. They are homoplastic or analogous and not homologous (see Chapter 7). Parallelism does this, too, and its results are even more difficult to distinguish from homology, but close parallelism is in itself evidence of relationship. Some of the major mistakes in interpretation of phylogeny, and on this basis in classification, have been caused by confusion of convergence with homology. These mistakes have nevertheless been detected and corrected. Undoubtedly we are still making some errors of this sort, but with increased attention to convergence and increased knowledge of its extent and character we can believe that such errors are no longer blatant. Most of them, surely, can be straightened out with further study.

In all cases of convergence, as in parallelism, there are three sorts of characters, which can usually be separated with adequate materials and intelligent study. First, there are characters that differ in the convergent groups and that were separately inherited from their different ancestries. These are the indications of their true relationships. There are, second, characters different in the two or more groups and special to them, not derived from their ancestries. Usually these characters can be related to differences in detailed habits, or ecological roles, between the two groups. Finally there are the characters that are convergent. These can usually be related to resemblances in habit and role—a very important point which will be mentioned again in a page or two.

The third possible relationship among separately evolving lineages, in addition to parallelism and convergence, is divergence: the groups become less similar as time goes on. All separate lineages diverge to some extent, even if they are in other respects parallel or convergent. Divergence is universal among separate lineages. Parallelism and convergence are not universal and are never complete; there is no known case in which two groups evolved in exactly the same way in *all* respects. Divergence is so common and so obvious that it hardly needs to be exemplified. Look about you at any two forms of life: you and your dog, grass and trees, rose and delphinium. The differences between them are the sequel of divergent evolution at some time in the vast history of life. By the way, the extreme diversity of life and the complete generality of divergence in evolution is sufficient evidence that undeviating trends are not universal or, one might properly conclude, even common in evolution. Every divergence means either change without trend or a change in trends.

Parallelism, convergence, and divergence are simply names for the three things that can happen among separately changing groups of organisms: they can remain about equally similar, become more similar, or become less similar. They describe but do not explain. To say that certain resemblances are "due to" or "caused by" convergent evolution is to some extent illogical. Even so, the statement is not meaningless and is often a convenient way of expressing the observation or opinion that the resemblances are *not* indications of phylogenetic relationships. We are left with the problem, which is basic to all evolutionary theory, of how such relationships do, in fact, develop.

In parallelism there are two different factors. The phenomenon occurs among similar animals, which are similar because they have about equally similar systems of development from egg, seed, spore, or whatever may be the means of origin of individuals in the group concerned. The similar systems of development depend, in turn, on similar heredity. What is inherited is an intricately coordinated, very complex chemical and physical mechanism. When these mechanisms are similar it is much more likely that similar than that different changes will arise in them. That is the agreed observation of geneticists in recent organisms. It is confirmed by the fossil record: similar changes more commonly occur and tend to be more closely similar in proportion to nearness of relationship of the organisms involved. That is the special factor in parallelism. A second factor is shared with convergence and divergence.

The special factor of parallelism cannot help to explain convergence and divergence. On the contrary, in convergent evolution similarities arise in spite of the fact that the developmental and genetic systems are different. In divergence differences evolve in spite of similarities in those systems.

The major clue to this puzzle is that parallel, convergent, and divergent characters can usually be related to the way of life of the organisms concerned. They are as a rule adaptive, that is, they are useful to the organisms in connection with their habits and habitats. Similar characters arising by parallelism or convergence are generally associated with similarity of adaptation and different characters arising by divergence with differences in adaptation. The remarkably convergent brachiopods, pelecypods, and corals clearly had similar adaptations to habit and habitat. All lived attached to the bottom in clear, shallow seas and obtained through the opening at the top of the cone small to microscopic animal food suspended in the surrounding water. The sabers of *Thylacosmilus* and the true sabertooths were certainly useful and used in the same way in obtaining their large, tough-skinned animal food. The other special resemblances between those two groups are all mechanically related to the functioning of the sabers. Resemblances between ichthyosaurs and cetaceans are obviously adaptive to pelagic, marine life.

That differences are also usually adaptive in nature is a very general observation. Sometimes this is obvious and simple. No one can doubt, for instance, that the striking differences between wild dogs and cats in structure, habits, and temperament are related to differences in prey, in methods of hunting, in social relationships, in care of the young, or in some other useful, hence adaptive, features of their separate ways of life. Other cases may be quite subtle. For instance, it has been found that dark coloration in some animals is not, in itself, of any particular adaptive significance but is an incidental result of physiological processes that are definitely adaptive. There are still many characters of plants and animals the adaptive value of which is unknown. It remains probable that some, of a rather minor and indifferent nature, do not have adaptive significance. Nevertheless when recent plants and animals are really well known as to environment, habits, and physiology it is generally found that their distinctive characters are adaptive.

The same conclusion can be applied to parallel, convergent, and divergent series of fossils, in part by direct observation and in part by again using the present as a key to the past. For instance, in the brachiopod-pelecypod-coral convergence direct observation of the fossils shows that they were inverted cones and lived with the apex of the cone attached to the bottom; observation of their fossil associations and the rocks in which they occur shows that they lived in clear, shallow seas; and inference from their resemblances to recent relatives shows that they lived on small animal food suspended in the water. In general, it is probable that characters involved in parallel and divergent evolution are usually adaptive, although it remains possible that a few are not. It is probable that characters involved in clear-cut cases of convergence are always adaptive. "Always" is a big word, and of course it cannot really be proved that there are *no* exceptions. At any rate, I know of no cases of convergence in which interpretation of the similarities as nonadaptive seems reasonable to me. This is a widespread opinion, although there are a few dissenters as there always are on such points.

Let us now revert briefly to trends and consider their relationship to adaptation. Trends certainly show that there is a factor that sometimes guides evolution. If evolution went on quite at random it would be impossible for it so often to continue in the same direction over long periods of time. Ever since the days of Darwin, whose theory of evolution was published in 1858, nearly a century ago now, it has seemed to many students that the nonrandom element in evolution is adaptation. (Indeed, this idea was already being occasionally expressed a century or more before Darwin, although as a speculation more than as a conclusion from evidence.) Other students countered that this could not be so, because trends keep right on going, "orthogenetically," and finally go to such lengths that they cause extinction. Those students were flatly wrong in their reading of the fossil record. They were reading their ideas into it and not reading what it clearly says. Trends as they really occurred, with all their irregularities and changes of rate and direction and occasional reversals, were adaptive. Today that is one of the most firmly established and most important inferences from the record of the history of life.

It has been mentioned that trends can become reversed and an example has been given: some lineages of horses became first larger and then smaller in long-range trends. In one such lineage, for instance, there was a decidedly fluctuating but fairly continual trend toward larger size for about 30 million years and then a better-defined trend toward smaller size for about 10 million years. This is an apparent contradiction of a well-known conclusion, first reached by paleontologists and supported by most of them in one form or another for the past fifty years: evolution is irreversible. That dictum is true, but it requires a little common sense in its application. By now I hope you are persuaded that paleontology is based on common sense and that many of its remaining doubts disappear when that homely commodity is applied. It is not common sense to conclude, as some enthusiasts formerly did, that a large animal cannot be the ancestor of a smaller one, that an enlarged tooth cannot later be reduced in size, or an attached animal cannot become free-living again. There are unmistakable, irrefutable examples of all three of those events and of many others like them in the fossil record.

Evolution does not repeat itself. The horses that become smaller were very different animals from their more remote ancestors of the same size. Free-living crinoids, which are descended from sessile forms and are, indeed, briefly sessile in their youth, are also very different from the ancient, free-living ancestors of all the crinoids. Whales returned to the water, ancestral home of all of us, but they did not become fishes.

It is more broadly significant to say not only that evolution is irreversible but that it is irrevocable. Organisms never return to an ancestral condition, and they also never lose the effects of that ancestry. The fact that parallelism and convergence never result in identity indicates that different ancestors do not produce identical descendants. It follows that if any ancestor of an existing form of life had been appreciably different, the existing descendant would have been different. The fossil record shows that there is no return to the past and no repeal of its effects.

The phenomenon of homology as a whole exemplifies the irrevocability of evolution. This is most dramatically evident in instances of transformation, radical changes of function richly exemplified by fossils and including some of the most extraordinary ways of organic change. New organs and structures do not arise on demand, nor do they simply appear as if by magic. They evolve from what is already there, by transformation within the existing developmental system and, again, by gradual change in sequences of varying populations. The necessity to build the new out of the old is the principal reason why evolution does not always seem particularly efficient to human eyes.

Any good engineer should be able to devise a better walking instrument than the leg of an amphibian. But the legs of amphibians evolved from structures, the paired fins of fishes, that originally had quite a different function: balancers and elevators for animals freely moving in water. By later, further transformation the limbs reverted to essentially their original function in ichthyosaurs and whales. The intermediate structural stages were irrevocable, and the paddles in those forms are quite unlike the fins of fishes even

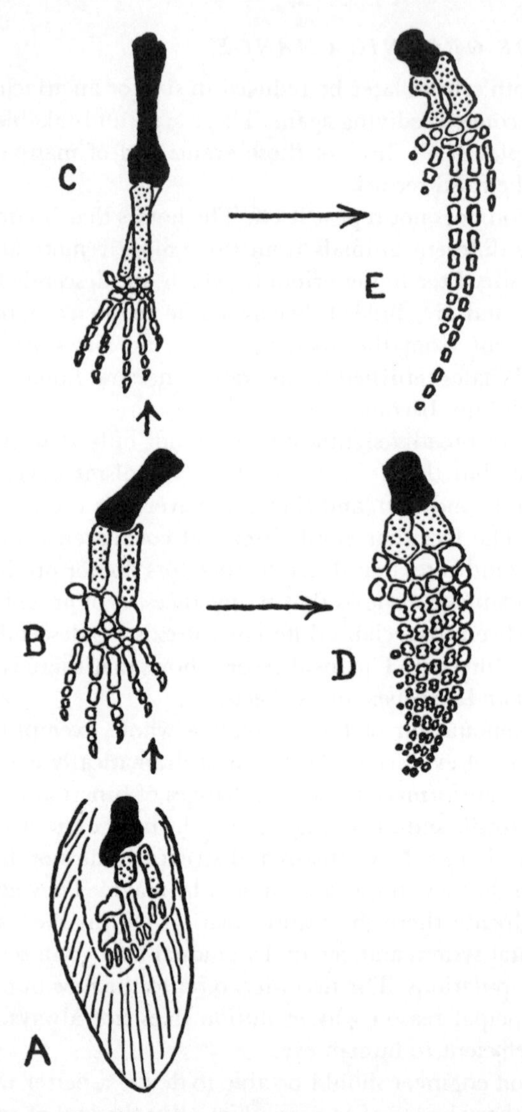

Fig. 42. The irrevocability of evolution. Evolution of the fore limb in some vertebrates. The effect of ancestral conditions cannot be lost: although the limbs are radically transformed in function and structure, each is built from the materials present in the ancestor. The bones in black (humerus) and stippled (radius and ulna) are homologous throughout, as are some others not specially indicated. An ancestral condition cannot be regained: D and E are flippers functionally similar

though they operate in much the same way. (Fig. 42.) Transformation of a very different sort led three times—in pterodactyls, birds, and bats—to the evolution of wings from walking legs. Three different ancestries were involved. *Their* special characters were irrevocable, too, and the three sorts of wings are quite different even though all serve well for flying. Throughout all these and many other divergent changes the old fish-fin structure was never wholly lost. The upper arm bone (humerus), for instance, is clearly homologous in all these later forms and is derived from a homologous bone in the fin of the now very remote ancestral fish.

This short account of the ways of change as seen in fossils should include some notice of two other, disparate topics: rates of evolution, and survival and extinction. Both are important and very complicated subjects which I have discussed in *The Meaning of Evolution* and *Major Features of Evolution.*

The most striking things about rates of evolution are that they vary enormously and that the fastest of them seem very slow to humans (including paleontologists, I may say). If any one line of phylogeny is followed in the fossil record it is always found that different characters and parts evolve at quite different rates, and it is generally found that no one part evolves for long at the same rate. The horse brain evolved rapidly while the rest of the body was changing very little. Evolution of the brain was much more rapid during one relatively short span than at any other time. Evolution of the feet was practically at a standstill most of the time during horse evolution, but three times there were relatively rapid changes in foot mechanism.

Rates of evolution also vary greatly from one lineage to another,

---

to each other and to the remotely ancestral fish fin, A, but they are quite different from A. They are also different from each other because of different ancestry (B and C, respectively); but they resemble each other more than they resemble A and to the degree that B and C are similar. A, fin of a fish of the type from which land vertebrates were derived (schematic, similar to *Eusthenopteron*). B, primitive reptile (schematic, similar to *Ophiacodon*). C, primitive mammal (schematic, similar to *Didelphis*). D, early ichthyosaur (essentially *Mixosaurus*). E, whale (essentially *Globicephala*). (Compiled from many sources, including Flower, Gregory, Repossi, Romer.)

even among related lines. There are a number of animals living today that have changed very little for very long periods of time: a little brachiopod called *Lingula,* in some 400 million years; *Limulus,* the horseshoe "crab"—really more of a scorpion than a crab—in 175 million or more; *Sphenodon,* a lizard-like reptile now confined to New Zealand, in about 150 million years; *Didelphis,* the American opossum, in a good 75 million years. These and the other animals for which evolution essentially stopped long ago all have relatives that evolved at usual or even at relatively fast rates. There are, further, characteristic differences of rates in different groups. Most land animals have evolved faster than most sea animals—a generalization not contradicted by the fact that *some* sea animals have evolved much faster than *some* land animals.

Without going into much technical detail it is impossible to be very specific as to how rapid are the fastest rates of evolution. Living populations have been seen to change appreciably in a dozen years or less, but this was under strongly exceptional conditions and the change was insignificant in comparison with the whole scope of evolution, less than the difference from one species to another closely similar. The rise of a new species in nature seems usually to have taken 500 thousand years or more. In the direct ancestry of our horse, a rather rapidly evolving line, the average time involved in change from one genus to another was well over 5 million years. Evolution has been a slow process; but it has been going on for perhaps 2000 million years.

To turn to the subject of extinction, four broadly general theories have been advanced: 1) a group of organisms has a definite span of life, as does an individual, and becomes extinct from old age; 2) evolutionary trends get started and cannot stop, so that groups become extinct when their trends go too far; 3) the environment kills off groups of organisms; 4) organisms lose adaptation, usually because their populations do not adapt to changes. The first two theories are historical curiosities and can be discarded without argument. No respectable evidence now supports them. The fourth theory, which also embodies the third and makes it still more general and more realistic, is undoubtedly correct. Extinction, like much else in evolution, is a two-way interaction. Changes constantly occur in the environment, which, broadly speaking, includes not only the physical surroundings of a group of organisms but also all other associated organisms and the group itself. Usually the group is adjusted to such changes or makes a new adjustment by evolutionary change. When it fails to do so it becomes extinct.

Sometimes it is possible to designate a more particular cause for extinction. This is especially true in cases where the change involved was the invasion of a competing form of life. More often no very precise cause can be assigned. No one knows exactly why dinosaurs and a host of other ancient forms became extinct. This is not because there is anything mysterious or metaphysical about extinction or because possible causes are unknown. It is just because there are many reasonable possibilities, and the record does not enable us to say in the particular case which of them were actually involved. All we can say at present is that something changed and dinosaurs did not. What changed and why the dinosaurs failed to cope with the change are among the things still hidden from us.

10. Theories of Evolution

EARLY IN the 19th century it was well known that fossils occur in a definite sequence and characterize different periods in the history of the earth. The idea had been suggested long before, and after the work of Smith, Brogniart, and Brocchi (see Chapter 2) no one in a position to judge seriously doubted it. The idea of evolution was also familiar to all the scientists of that day, but its acceptance was by no means so general. It now seems to us obvious that the two ideas reinforce and complement each other: the fossil sequence is the record of evolution. Yet this connection was not accepted by the students of fossils in that period. Almost to a man they denied the truth of evolution, and they cited the fossil record as evidence for their stand. Knowledge of that record was still very incomplete, and we have seen that the record itself is incomplete. What the early paleontologists and geologists thought they saw was a sequence of quite distinct faunas and floras. They had not yet found or they overlooked evidence of evolutionary transition from one stage to the next.

The most distinguished paleontologist of the time was Cuvier (1769–1832). He is, indeed, often hailed as founder of the science of paleontology, although he had many predecessors in the study of fossils and was professor of anatomy, not of paleontology, at the National Museum of Natural History in Paris. (A detailed study of the history of any field of science gives the impression that no specialties and no theories appear full blown or have a precisely defined time of origin.) It was Cuvier's view that each successive fauna, as known to him, was the result of a repeopling of the earth after a great catastrophe that had wiped out many previous species. The last of these catastrophes was of course the Biblical deluge. As to where new species came from after a catastrophe, Cuvier was quite positive that they did not evolve from older species, but otherwise he hedged. He was inclined to believe that they had existed all along and that when they appeared in the fossil record "they must have come from elsewhere." This is, indeed, commonly true of new groups that appear suddenly in the record, but as a general explanation of the origin of new species it seems curiously evasive.

Alcide d'Orbigny (1802–57), first professor of paleontology at the Paris museum, squarely faced the problem and came up with an answer even further from the truth. He taught that life on the earth was completely wiped out in each catastrophe and an entirely new set of species specially created in each successive stage.

Cuvier was an unusual combination of spellbinder and scientist. When his evidence was thin he made up for it by oratory. He produced a great body of work still valid and valuable today, but he also effectively silenced opposition to his entirely wrong theory of catastrophes or revolutions, and he retarded acceptance of the truth of evolution. Almost all students of fossils, of whom there were many, during the first half of the 19th century accepted his authority and followed his views. Most of them also accepted d'Orbigny's modification of Cuvierian theory. "Successive creations" were for a time the easy way out of increasing embarrassment as they came to see that the fossil record *does* show a progressive development of life not accounted for by the Book of Genesis.

In the meantime and even at the institution of Cuvier and d'Orbigny, heretical views were being voiced by nonpaleontologists. Lamarck (Jean-Baptiste Demonet, Chevalier de Lamarck, 1744–1829, to give him his full style) was professor of invertebrate zoology there. Particularly in his *Zoological Philosophy,* published in 1809, he came out wholeheartedly for evolution as the general explanation of the history of life. His associate in vertebrate zoology, Étienne Geoffroy Saint-Hilaire (1772–1844), accepted this view, with differences of opinion as to details of the process. Cuvier scorned to reply publicly to Lamarck's arguments, but he debated Saint-Hilaire at the Academy in 1830 and scored such an oratorical victory that little more was heard of evolution in France for another generation.

Lamarck's literary style was not brilliant, and his remarks on anatomy and physiology include much that was even then recognizable as nonsense. This helps to explain why his influence on his contemporaries was virtually nil. He was nevertheless the first really important figure in the development of modern evolutionary theory. His particular theory of how evolution occurs was, however, quite different from that later called "neo-Lamarckian." He believed, first of all, that there is some mysterious, inherent tendency for life to progress from the simple to the complex, from the less to the more perfect. This is a very old idea, foreshadowed by Aristotle. It is now known to be incorrect, and yet it became so confused with the whole concept of evolution that it still exerts a sort of vestigial, hidden effect on some students of the subject.

Lamarck was acute enough to observe that life does not really form such a progression. He explained away this inconvenient fact by saying that the true course of evolution is perturbed by local adaptations. Adaptation was said to result from the activities and habits of organisms, which modify their anatomy. He assumed, as did almost everyone from the dawn of history down to and including Darwin, that such modifications would be inherited in like form by offspring.

The place of Charles Darwin (1809–82) in the history of evolutionary theory is known to everyone, although not always quite correctly known. It is, again, typical of the history of science that there is practically nothing in Darwin's theories that had not been expressed by others long before him. His predecessors, however, were long on speculation and short on facts, and much that they said impressed their contemporaries as silly—as, in most cases, it does us today. Darwin brought together an enormous body of solid, pertinent fact, he reduced speculation to a minimum, and nothing he wrote (even though some of it later proved to be quite wrong) could be characterized as nonsense. He demonstrated to the satisfaction of the whole scientific world that evolution has, in fact, occurred. He also produced a particular theory as to how it occurred, of which more later.

From our present point of view an especially interesting thing about Darwin's principal work, *The Origin of Species* (1859), is that it devoted two chapters to explaining why paleontology and paleontologists up to that time did *not* support the truth of evolution. The most objective proof of that truth was to come from the fossil record, but it was clear that further search and study from this point of view were necessary. Some of the old-timers, such as Owen or Agassiz, were unable to adjust to this revolution in thought. Almost immediately after publication of *The Origin of Species*, however, a large number of evolutionary paleontologists was at work. Among them were T. H. Huxley (1825–95) in England,

Cope (1840–97) and Marsh (1831–99) in the United States, Gaudry (1827–1908) in France, Kovalevsky (1842–83) in Russia, and Rütimeyer (1825–95) in Switzerland, to mention only a few.

This renewed work rapidly proved that evolution is a fact. Accumulated series of ancestors and descendants and discovery of numerous transitional forms among fossils left no reasonable doubt. Achievement of adequate proof was gradual of course, but two landmarks may be mentioned: Marsh's completed demonstration of the essential stages in the evolution of the horse (1879), and the review of the evolution of all groups of fossils then known in volumes of the great handbook by Karl von Zittel (1839–1904) of Munich, beginning publication in 1876.

Adequate proof that evolution did occur was a first necessity and a great achievement. Much more confirmation has piled up since, but more was hardly needed after about 1880. In the meantime attention was directed to the next and more difficult problem: how and why has evolution occurred? Lamarck had made a rather primitive attempt to cope with this question, and Darwin had made a much more substantial but still incomplete contribution to its solution. In the later years of the 19th century and early in the 20th many different views were aired, with paleontologists taking prominent parts in the arguments. Some of the theories ascribed evolution vaguely to a life force of some kind or to other virtually unknowable metaphysical factors. Such theories are nonexplanatory and stultifying. They also seem to me, and to most other students of the question, quite inconsistent with the fossil record, and no further discussion of them is needed here.

Two opposing schools of theory were based realistically and naturalistically on the material facts of life and its record: neo-Lamarckism and Darwinism. The so-called neo-Lamarckian school was only in part derived from Lamarck and, indeed, it conflicted with much of his doctrine. It contended that materials for evolution were individual modifications caused by the reactions of organisms (a point really Lamarckian) and by action of the environment on organisms (a point flatly denied by Lamarck). It necessarily insisted that such modifications were heritable; otherwise they could have no direct influence on evolution. (Lamarck believed this, but so did Darwin and most other students from antiquity to about 1900.) It omitted Lamarck's main thesis: that of an inherent progression in evolution. That idea was taken over by some of the metaphysical theories.

Neo-Lamarckism stressed interaction of organism and environment. Adaptation was its keynote, and it professed to explain adaptation in a particularly direct and simple way. Individuals adapt themselves, and, said the neo-Lamarckians, that is all there is to it. Paleontologists were then, as they are still, especially impressed by adaptation and by its slow and progressive development through time. They were not then (but they are now) particularly concerned with the actual mechanism of heredity and so did not observe that this was a crucial difficulty for neo-Lamarckism. Many paleontologists were neo-Lamarckians, and that theory seemed for a time to draw substantial support from the fossil record. The work of Cope, an exceptionally able paleontological theoretician, is an example.

Even if true, neo-Lamarckism could not be a general explanation of evolution. Many evolutionary events known positively to have occurred could not conceivably be explained in this way. For instance, the nonreproducing castes of insects cannot have inherited their characteristics from ancestors in which those same characteristics arose as modifications. Neo-Lamarckism finally came to an end with definite establishment of the fact that the sort of inheritance required by that theory cannot occur. There are still a handful of neo-Lamarckians in countries where known truths, possibilities, and known errors may be expressed with equal freedom. The only significant support for neo-Lamarckism now, however, is in the U.S.S.R., where *only* error may be expressed if the political bosses so decree. They have decreed that Soviet biologists must be "Michurinists." Michurinism, put over by Lysenko, is a reactionary form of neo-Lamarckism. It should be added that before this dictatorial action Soviet scientists were making excellent and substantial contributions to modern evolutionary theory.

Darwin himself accepted what was later called neo-Lamarckism as a subsidiary factor in evolution. His followers, the neo-Darwinians, rejected it and concentrated attention on what Darwin had designated as the main factor: natural selection. In natural populations it is usual for more young to be born than can survive to reproduce in turn. On the whole, those that do survive are better fitted to their conditions of life. Since some, at least, of the characteristics making them more fit are hereditary, changes in heredity of the group through the generations are in the direction of greater fitness. This factor, inherently so reasonable and probable, would evidently tend to produce progressive adaptation. Ever since its first clear and wholly logical formulation by Darwin (and Wallace, 1858, and more extensively by Darwin alone in 1859), many students have accepted it as the explanation of adaptation or of evolutionary change in general.

There were nevertheless strong objections to the theory of natural selection, in which many paleontologists joined. Experimental proof of the operation of natural selection was slow in being achieved. It has now been amply demonstrated, as well as the actually observed operation of natural selection in nature. A whole series of counterarguments was based on judgment that distinctions so slight as to be ineffective for natural selection were nevertheless involved in evolution. These arguments have also been entirely controverted in more recent years. It has been demonstrated that natural selection is much more subtle and powerful than at first appeared. Any variation observable by us is under favorable circumstances sufficient for the action of natural selection. Another series of arguments was based on the claim that many changes in evolution are nonadaptive and would not be favored or might even be opposed by natural selection. The supposed examples are complex, but in general they have turned out to have one of three explanations. In some cases the claimed phenomena are not real, for instance those of nonadaptive orthogenesis (see Chapter 9). In others the changes were really adaptive or may most reasonably be so regarded; adaptation is an extremely intricate process and human judgment of what is or is not adaptive is often fallacious. Finally, it is entirely possible that some truly nonadaptive change does occur, because natural selection is not necessarily effective under all conditions. That natural selection causes adaptation is not at all controverted if adaptation is found not to be perfect or universal.

Most of the objections to natural selection that formerly loomed large and that long spurred a search for alternative explanations have thus been fully removed. To that extent the original Darwinian theory has been substantiated. There were, however, two much more serious objections, and these have led to essential modifications in the theory. In its original form the theory took the existence of heritable variations more or less for granted. It therefore was very incomplete in not accounting for the *origin* of such variation. Darwin did attempt to account for this but failed. A related problem is that Darwinian natural selection seems to explain only the elimination of the unfit and not really the rise of the fit. The existence of a third and at least equally serious problem was hardly realized until it was solved: Darwin assumed that inheritance is blending, that a large and a small parent, for instance, would always have offspring of intermediate size. If that were true progressive change by natural selection could not occur. At about the time when this objection became apparent, it was found that inheritance does not blend in this way.

Around 1900 and thereafter geneticists began to learn just how variations do arise and are inherited. For present purposes the essential point is that inheritance is dominated by a developmental system controlled by definite structural and chemical units in the reproductive cells. Those units are the chromosomes. From time to time the units undergo changes of various sorts: increase or decrease in number, changes in arrangement, or chemical changes within the chromosomes among the still smaller units, the genes, that occur in them. Such changes, called in general mutations, produce through the developmental system new characteristics in the organism as a whole. The result is the rise of characteristics not inherited from the parents and yet heritable by the offspring. That is the ultimate source of hereditary variation.

When these facts were being discovered some geneticists at first believed that mutation was the whole story of evolution and that natural selection, along with neo-Lamarckism, was supplanted. It seemed to them that new sorts of plants and animals simply arose by mutation and that selection had no role beyond the gross fact that the new organisms must be capable of survival. It is a peculiarity of mutations that they show no special tendency to occur in the direction of past or current evolutionary change or of increased adaptation and are to that extent random. For the mutationists, evolution as a whole was therefore an essentially random process.

Paleontologists knew that the early mutationist theory could not possibly be true. Some of the evidence was summarized here in Chapter 9. Trends may continue slowly in the same direction for millions of years. Most of the changes observed in fossils are clearly nonrandom and adaptive throughout. Abrupt origin of new groups by mutation is not substantiated by the fossil record, while gradual change in varying populations is at least common. The known features of the history of life certainly cannot be accounted for by random mutation alone.

This radical disagreement had disastrous effects on the study of evolutionary theory, although fortunately they did not long endure. Some students despaired of finding any explanation for evolution, or turned again to metaphysical pseudo explanations that did not really explain anything. On the whole, antagonism developed between geneticists and paleontologists, along with many neontologists. For a time each side was so sure the other was wrong that they went their own ways without consideration of the whole picture of which each saw a part. Yet along with their separate theories, which were incomplete and partly wrong on each side, each had quite incontrovertible facts. That the facts seemed to conflict was the fault of the students and their theories, not of the facts, after all.

It is the great achievement of the present generation of students of evolution that the conflict has been fully resolved. A theory has been developed that takes into account the pertinent facts of paleontology, neontology, and genetics and that is consistent with all. Because the theory involves natural selection as an essential factor it is sometimes called neo-Darwinian. That is something of a misnomer. As between Darwin and Lamarck, the theory owes much to Darwin and little or nothing to Lamarck. Nevertheless its genetical side, which is at least as important as selection, was wholly lacking or wrong in Darwin's own theory, and even selection is given a broader and somewhat different meaning from Darwin's. Since the theory is a synthesis from many forerunners and from many fields of biological science, it is often and less misleadingly called the synthetic theory of evolution or the modern evolutionary synthesis.

Early geneticists—in the young science of genetics "early" means roughly from 1900 to 1930—were necessarily concerned mainly with particular mutations and the inheritance of these by individuals. Such an approach is similar to that of the old typological systematics (see Chapter 7), which studied single characters and their combinations in "types" as abstractions and did not consider their real rise, variation, and change in populations. The first abortive attempts to reconcile paleontology and genetics were made on this basis. New types were considered as mutant forms arising at one jump and as such in individuals. We have seen that this view is really inconsistent with much that is known from the fossil record. It is also inconsistent with the most reasonable interpretations of living populations in nature. Finally it turned out also to be inconsistent with the findings of genetics as that science matured.

Truly fruitful synthesis could be achieved only when both genetics and paleontology advanced beyond the typological stage. Beginning in about 1930 and in full swing today has been the development of population genetics. More or less simultaneous has been the development of what might well be called population paleontology and population systematics. It is through these movements that the varied approaches to the problem of evolution through paleontology, systematics, and genetics have turned out really to lead to the same result.

The core of the synthesis is not particularly complex or esoteric. New organic characters, variants and new structures, arise by mutations in populations. The frequencies of particular characteristics and, what is even more important, of their combinations result not only from mutation but to even greater extent from processes involved in the reproduction of the population. In sexual reproduction there is a constant shuffling of genetic combinations. Certain changes in frequencies of mutations and their combinations occur at random with respect to adaptation. (That they are in this sense "random" does not mean that they have no determinable material causes.) Other changes are systematic. There may be a consistent tendency in reproduction for offspring in each generation to have more of certain mutations and combinations, less of others, than the last generation. From one generation to the next the changes in such frequencies are usually very slight, or indeed practically indistinguishable from small random fluctuations. In the long run, however, the cumulative effect becomes quite appreciable—evolution is indeed a slow process as the fossil record shows. This consistent differential in reproduction is what is meant by "selection" in the modern theory. If adaptation is understood in a broad sense, the differential tends always in the direction of adaptation.

Darwinian natural selection, death or survival of certain sorts of individuals, may and usually does lead to reproductive or genetical selection. When it does not, it has no real effect on evolution. When it does, it is a special case of genetical selection. Genetical selection is more general and does not necessarily involve Darwinian natural selection. To take only one simple case, it is evident that two individuals that survive equally long may nevertheless have quite different numbers of offspring. Then genetical but not Darwinian selection has occurred.

Genetical selection meets the old objection that Darwin's natural selection was not "creative," that it eliminates and does not originate. Genetical selection determines what mutations will, in fact, spread in populations and how they will be combined. This is decidedly a creative role. In much the same sense an architect is creative. He does not make building materials, but he determines what materials shall be used and how they shall be put together to produce an organized result.

In populations that do not reproduce sexually the basic situation is even simpler. For the most part their evolution is an interplay of mutation and Darwinian selection. Organisms reproducing asexually are less common than those reproducing sexually and have played a lesser role in progressive evolution. Sexual reproduction must have arisen very early in the history of life, and it now occurs even in very lowly organisms. It has recently been found that in some, at least, of the bacteria a form of sexual reproduction occurs, and this is widespread and well known in many other protistans (see the appendix).

There are numerous different factors involved in these processes. Each has many variations of kind, intensity, and direction. Their interactions are extremely complex, and the study of particular phases and aspects of evolution is almost incredibly intricate. Yet all involve the relatively simple basic processes that have now been summarized. In detailed, technical studies it has been established that these processes are not only consistent with the fossil record but are also adequate to explain it.

You will have noticed that this explanation is complete at a certain level or up to a certain point but that it still leaves deeper problems unsolved. Most importantly, it does not explain why mutations arise or why and how they produce their particular effects. These problems have not yet been solved, although progress is being made and there is every reason to think that they are soluble. It seems unlikely that their solution will have much effect on current interpretations of the fossil record, even though it will surely deepen our understanding of the whole history of life.

11. Fossils and Mankind

SOONER or later every student of fossils is asked, "What good is it? Your subject may be interesting to you as a profession and to me as a hobby, but after all what contribution does it really make to society?" The question is fair and it requires a plain answer. It should be asked of and answered by all scientists—as well as all businessmen, politicians, labor leaders, housewives, and indeed everyone. The paleontologists' answer is largely implicit in previous chapters. This book may fittingly close with some more explicit consideration of the human value of fossils and their impact on mankind.

What the questioner often has in mind is monetary value. In our commercial countries there are many people who think of themselves as "hardheaded." Production of something that can be sold for money appeals to them as practical and proper. Anything that does not have obvious cash value is likely to be considered a pastime unworthy of occupying the full attention of adult and sober men. Even these incomplete souls can be shown that fossils are valuable. As was seen in Chapter 3, they are indispensable aids in the discovery and exploitation of mineral wealth, especially of petroleum but also to some degree of any resources associated with sedimentary rocks. Many paleontologists are paid good salaries by commercial corporations and more than earn their salaries in terms of cash return. Moreover, the effectiveness of their work depends on the researches of "pure" paleontologists, who have no immediate concern with industry. Even such studies, apparently unprofitable in terms of money, as those of ancient communities (Chapter 5) turn out to have cash value when applied to economic paleontology.

These economic values would be enough to justify the study of fossils and the lives of the many men who have devoted themselves to this subject. They are, however, far exceeded in real importance by less tangible values. Among these are many values that may be called aesthetic. By this I do not mean only that fossils have beauty, but more broadly that they can and do give emotional and intellectual pleasure comparable to that aroused by any fine work of art or nature. The living organisms of the past add to this special aesthetic aspects of their own.

Many fossils are literally beautiful and appeal to the artist as well as to the paleontologist. The minute, complex perfection of a foram, the exquisite symmetry of a brachiopod, or the intricate tracery of ammonite sutures is beautiful in the same sense that a Greek statuette or a Dürer engraving is beautiful. Other fossils, such as fossil bones, are rarely beautiful in quite that sense. They gather beauty when considered as functional structures, elegantly adapted to the motions and stresses of their use. They are to functional modern design what a brachiopod is to an old master. Moreover, like any fossils they embody the beauty inherent in life, and the mystery of life and of the great past. To anyone who has studied them all fossils have real beauty—a perception sometimes startling to the uninitiated when applied to what is to them a chunk of dull rock.

Even the uninitiated do find emotional values in fossils. The dullest visitors to museums display some sense of awe when they first stand before a dinosaur skeleton and realize that this great frame lived and breathed sometime in the dim reaches of the past. The thought that little eohippus became a horse is at least diverting, and some spiny reptiles from the Permian are as comical as a cartoonist's creation. Indeed prehistoric animals have inspired many a cartoon. Perhaps those brave enough to define relative values would not put amusement at the top of the list, but the fact is that fossils are often amusing. This, too, is a value not to be scorned, and one that is not scorned by those whose interest in fossils is otherwise most serious. Museums are for entertainment as well as for instruction.

With only a little more knowledge—such knowledge as I hope this book has now given—more profound values are acquired. One of these is an expansion of personal experience and life. Our allotted span is a few score years, and most of us can see with our own eyes only a minute part of the earth around us. But our minds need not be restricted to these narrow limits of time and space. They can range throughout the past and can see all the curious creatures and scenes of life's history in ever changing sequence. A fair title for a book on paleontology might be *How to Live a Billion Years.*

Paleontology is the science that reveals the course of evolution, and it is one of the sciences concerned with the factors and causes of evolution. This has aspects recognizable as "practical" even to the hardheaded. Evolution is the process by which plants and animals change. Knowledge of this process is the most effective and practical way to speed up changes and to guide them into channels useful and profitable to man. These possibilities are already beginning to be realized. Possibilities for the future are far greater as we learn more about evolution and become more efficient in applying this knowledge.

Power over his environment is one of the distinctive features of man. Much of human prehistory and history can be written in terms of progress in achievement of this power. Now the most important part of our environment is not its physical but its biological aspects. Control over the world of life in which we live and of which we are a part is essential to fulfillment of our self-determined destiny. Further progress in this respect will require, indeed, cooperation of all the sciences, but less is to be expected from the physical sciences than from paleontology and the other sciences concerned directly with life.

False theories as to the development of life, such as the theory now officially imposed in the U.S.S.R., are certain in the long run to stultify efforts to utilize or to improve the forms of life, including man himself. Man has evolved in the past and surely will evolve, one way or another, in the future. Is that future all planned for us so that nothing we do about it means much? Or is change dependent on sudden random appearance of mutant "types," good or bad? Is it the cumulative result of environmental modifications? Or does it depend on differential reproduction of genetical systems? It is really clear already, and partly because of the evidence of the fossil record, that the latter is the true alternative. Neglect of this fact and reliance on any false theory of evolution can only eventuate in hopeless degeneration.

The long, long history of life is a marvel, but it is not a miracle. It is a slow interplay of material processes from which new configurations of matter and energy gradually emerge. At each stage some forms are arrested in their development. They cease to change essentially or they continue to change only in adaptation to minor modifications or subdivisions of the situations in which their ancestors lived. Others show cumulative change adapting them to new and different ways of life or improving the adaptation to ways already acquired. All along the line many forms of life are committed to adaptation that finally becomes disadvantageous, and they become extinct. In the long run extinction is the common lot, survival the exception.

One of the things most clearly conducive to progression with survival is increase in perception and in the power to react appropriately to what is perceived. Sense organs for receiving significant stimuli from the surrounding world become more specific and more effective. Nervous systems that correlate perceptions and that relate them to reactions arise and become more elaborate and more elastic. Of all the progressions seen in the history of life this most merits the name of progress, that is, not merely cumulative change but also change for the better.

The culmination of this sort of progress is man. The extreme elaboration of man's perception and reaction has, among others, two consequences of overwhelming importance. First, he has to a unique degree consciousness of the world and flexibility of reaction. He can learn new responses to stimuli and can, indeed must, choose between alternative responses. Second, he has developed social organization embracing the whole of his extremely varied and numerous population. In man new directions and new possibilities of progressive evolution emerge fully for the first time in the history of life. His is, indeed, a new sort of evolution, which does not replace but is added to the older, organic evolution. Man's unique sort of evolution is based on his flexibility of response, his ability to learn, and his social organization. That organization is not really at all like the old, instinctive social organizations of some insects. It has quite distinct bases and altogether different possibilities.

Our knowledge of all this, with all that it implies for man's present and future, is not, to be sure, the contribution of paleontology alone. But paleontology has materially contributed to it. Without the study of fossils our knowledge of these supremely important matters would be much less sure, less clear, and less extensive. Unfortunately there are still numerous men who really prefer not to know the truth if it disturbs an error that they have cherished.

With realization of this contribution of the study of fossils to man's understanding of man, only they can continue to ask "What good is it?"

The fossil record is the record of the process that has produced us along with all the other marvels of life. The actual record of man's ancestry is conclusive as to general trend, but very incomplete as to precise detail. Even from the point of view of understanding our own place in the universe it is not so much this specific point that is important: it is the whole interpretation of the record, whether of coal plants, of trilobites, or of apes.

As applied to mankind that interpretation shows that we did not appear all at once but by an almost incredibly long and slow progression. It shows, too, that there was no anticipation of man's coming. He responds to no plan and fulfills no supernal purpose. He stands alone in the universe, a unique product of a long, unconscious, impersonal, material process, with unique understanding and potentialities. These he owes to no one but himself, and it is to himself that he is responsible. He is not the creature of uncontrollable and undeterminable forces, but his own master. He can and must decide and manage his own destiny.

These conclusions, which embody the most profound of all the values of paleontology, cannot help but arouse emotion. In some the emotion is fear. In all the emotions should be pleasure in human self-reliance and determination for the proper exercise of man's responsibility to man.

APPENDIX. *A Review of the Forms of Life*

This summary is intended to show how varied life is and, in a general way, what the different forms are like. It includes only the broader groups: in terms of the Linnaean hierarchy (Chapter 7) the phyla, classes, and some orders. Most of the really important phyla and classes are included, but the review does not pretend to be complete even for these large groups. For instance, several small phyla of living animals of no great significance at the present time and with little or no fossil record have been omitted altogether. Groups well known as fossils are emphasized, and for each group the time of its first appearance is noted and usually something is said about the general nature of its record.

All students do not exactly agree as to the groups to be recognized or the names and ranks to be given them. I have followed either the consensus or a recognized recent authority in each case. For our present purpose the differences among specialists do not matter.

KINGDOM PROTISTA

These small organisms used to be called "one-celled animals and plants." Recently it has been seen that their bodies are not closely analogous to a single cell of other organisms and that they are better understood not as one-celled but as noncellular or not divided into separate cells. Also, at this level the distinction between plants and animals is not clear. Some of these forms could equally well be defined as one or the other, and some act like plants part of the time and like animals at other times. It is therefore becoming customary to place these noncellular forms in a separate kingdom, Protista, from unknown early members of which plants and animals, strictly speaking, were derived.

Class Sarcodina

Foraminifera. Small animal-like protistans, most of which form a shell by cementing sand grains or, especially, by secreting lime (calcium carbonate). Most of them are the size of grains of wheat or smaller, but a few reach diameters of several inches. The shells assume an astonishing variety of forms. They appear sparsely and doubtfully in the Cambrian, definitely in the Ordovician, and are abundant from the Devonian to the Recent. Forams are the most important microfossils and are essential aids in subsurface correlation (see Chapter 3). (See Fig. 44E, also Fig. 7.)

Radiolaria. Microscopic marine protistans most of which secrete an outer skeleton, a sort of lattice of silica. They are very abundant and varied today and are known as fossils from the Cambrian onward, but their history is not well worked out. Many of the textbooks still say that pre-Cambrian radiolarians are known, but this was based on what now seems to have been a series of mistakes.

Class Flagellata

These microscopic protistans, with one or more whiplike threads or flagella, are so fully intermediate between plants and animals in physiology that they used to be claimed by zoologists as Protozoa and by botanists as Algae. A few of them secrete hard parts, known sporadically from the Cambrian onward, but their fossil record is insignificant.

Class Bacillariales

Diatomaceae. Plantlike protistans that secrete skeletons of silica with extremely varied and beautiful geometric patterns. Occurrence in the Jurassic is perhaps doubtful, but they are certainly identified in the late Cretaceous and are common fossils in Cenozoic rocks. (See Fig. 43A.)

Class Bacteria

The essential role of bacteria in nature was mentioned in Chapter 5. There is abundant evidence that they have played this role since early in the pre-Cambrian, virtually throughout the history of life, but accurate identification of fossil bacteria is practically impossible. Many students still consider the bacteria to be Algae, but their activities are not fully plantlike and they are noncellular.

KINGDOM PLANTAE

Although some plants have secondarily lost this character, the fundamental characteristic of plants is the ability to use radiant energy to synthesize organic from inorganic chemical materials, the process of photosynthesis. Most true plants are made up of numerous separate cells, with some differentiation of the cells, and all are either photosynthetic or derived from photosynthetic ancestors.

Algae

"Algae" is a general term for plants divided into several (often five) distinct phyla in modern botanical classifications. All are relatively simple, primarily aquatic plants without sap-carrying vessels or highly specialized reproductive organs. Many of them secrete calcium carbonate, and such calcareous algae have been important reef- and rock-formers throughout the whole recorded history of life. Algal reefs from the middle pre-Cambrian or somewhat earlier are the oldest certain traces of life. Fossil algae frequently lack features used in distinguishing modern phyla and lesser groups. Those that can be precisely classified belong for the most part to the phyla Chlorophyta (green algae), with some recent families already present in the Ordovician, and Rhodophyta (red algae), a less abundant and apparently less ancient group.

Charophytes are bushy but simple aquatic plants sometimes included in the green algae and sometimes considered a separate phylum. They are of paleontological interest because their reproductive structures (oögonia) are rather common fossils from Devonian to Recent and sometimes make up whole beds of fresh-water limestone. (See Fig. 43B.)

Phylum Mycophyta

The mycophytes are the fungi: molds, yeasts, mushrooms, and their relatives. They were derived from one or perhaps several groups of algae with loss of photosynthesis. They have existed since the Devonian, at least, but fossil fungi are rare and usually too poorly preserved to have much interest.

Phylum Bryophyta

This group includes the liverworts and mosses. Their fossil record is very poor. Their earliest fossils, both liverworts and mosses, are Carboniferous.

Phylum Psilopsida

Psilophytales. This group is of special interest because it includes the oldest and most primitive vascular plants, that is, with specialized tissues for conducting fluids through stem and branches. They are known from middle Silurian to late Devonian. Specimens from Rhynie in Scotland are so well preserved that their anatomy is as well known as that of most recent plants. They were simply branched, without true roots and with no or small and simple leaves. They reproduced by spores. (See Fig. 43C.) Two peculiar plants still living, *Psilotum* and *Tmesipteris,* may be relicts of this ancient group.

Phylum Lycopsida

Lycopods are somewhat more complex spore-bearing, vascular plants, the smaller stems and branches of which are densely covered with simple

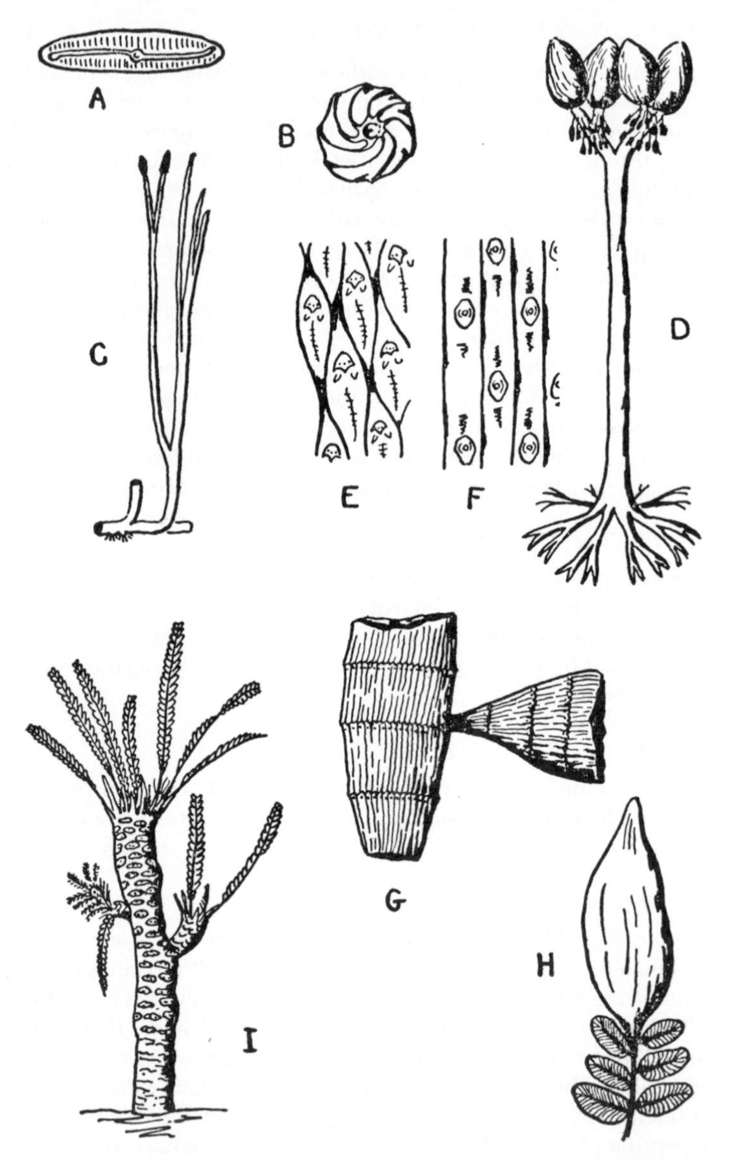

FIG. 43. Some fossil plants. A, a plantlike protistan, Class Bacillariales, *Navicula,* a Tertiary and Recent diatom, greatly enlarged (after Grabau). B, Charophyta, *Trochiliscus,* oögonium as preserved, Devonian, original about 1/32 inch in diameter (after Hirmer). C, Phylum Psilopsida, Psilophytales, *Rhynia,* restoration, Devonian, original about

foliage, and which have differentiated roots. They occur from late Silurian to Recent. The living relicts, four genera, are not uncommon and include hundreds of species. They are all small, but in the Carboniferous lycopods included large trees. Two of these, *Lepidodendron* and *Sigillaria,* were especially abundant and form much of the Carboniferous coal. (See Fig. 43D, E, F.)

Phylum Sphenopsida

These are still more advanced spore-bearing, vascular plants, readily distinguished from lycopods by their jointed stems with whorls of leaves around the joints (nodes). They are doubtfully known in the early Devonian and definitely from the middle Devonian to Recent. Like the lycopods, they became large trees in the Carboniferous, and the rather bamboo-like *Calamites* is abundant in coal deposits. (See Fig. 43G.) The living representatives are the small scouring rushes or horsetails, belonging to only one genus, *Equisetum,* but very widespread and often locally abundant.

Phylum Pteropsida

Following some recent students, especially C. A. Arnold, all the higher vascular plants, including the spore-bearing ferns and all the seed-bearers, are here put in a single division or phylum. In spite of their tremendous diversity they seem to have expanded from a single stock in the Devonian, when fundamental differentiation of land plants took place.

Subphylum Filicineae

These are the true ferns, distinguished from other pteropsids by the fact that they still reproduce by spores rather than by seeds. They have a long history, from the middle Devonian, at least, but one that has been

---

6 inches in height (after Kidston and Lang). D, Phylum Lycopsida, *Sigillaria,* restoration of whole plant, Carboniferous, original about 30 feet high (after Hirmer). E, bark of a related lycopsid, *Lepidodendron,* showing leaf scars (after Lesquereux). F, bark of *Sigillaria* (after Lesquereux). G, Phylum Sphenopsida, internal cast of pith cavity of *Calamites,* Carboniferous, including main stem and a branch, original diameter of main stem about 1½ inches (after Grand'Eury). H–I, Phylum Pteropsida. H, Pteridospermae, *Neuropteris,* seed and part of foliage, Carboniferous, original seed about 1¾ inches long (after Thomas and Edwards). I, Cycadophyta, *Williamsonia,* a Jurassic cycadeoid, restoration of whole plant, original trunk about 6 feet high (after Sahni).

difficult to disentangle because the leaves may look exactly like those of pteridosperms (below).

Subphylum Pteridospermae

Pteridosperms, now long extinct, were plants with fernlike foliage but reproducing by means of true seeds. Occurrence in the Devonian is doubtful. They are surely present from early Carboniferous to middle Jurassic. Small forms occurred, but large pteridosperm trees dominated the Carboniferous vegetation. These plants were long confused with ferns and are still usually called seed ferns, but it would be more nearly correct to think of them as gymnosperms (see below) with fernlike foliage. (See Fig. 43H.)

Subphylum Gymnospermae

Gymnosperms can be simply (but not very significantly) distinguished from pteridosperms by the fact that they do not have fernlike foliage, and from angiosperms (see below) by the facts that they do not have true flowers and that they bear the seeds exposed. (Gymnosperm means "naked seed.")

Cycadophyta. Living cycads are extensively cultivated and are familiar to most of us, although often confused with palms. They are very palmlike, but are gymnosperms while palms are rather advanced angiosperms. True cycads are known from the late Triassic to Recent, but are uncommon fossils. From late Triassic to late Cretaceous there were abundant plants, the cycadeoids, that looked almost exactly like cycads as to trunk and foliage but that bore the seeds in remarkable flower-like organs. These are, in fact, often called flowers, but their structure was quite different from that of a true flower and the resemblance is superficial. (See Fig. 43I.)

Ginkgoales. The living ginkgo tree is now also widely familiar in cultivation. It is a single relict species (*Ginkgo biloba*), preserved from extinction in China and mainly in cultivation, although possibly wild individuals have recently been reported there. The leaves are highly characteristic and superficially resemble those of the maidenhair fern. Ginkgos are known from the late Permian, and in the Mesozoic they occurred in considerable variety all over the world. (See Fig. 44C.)

Cordaitales. The cordaitales are extinct, Paleozoic (Devonian to Permian), gymnospermous trees with large, straplike leaves. *Cordaites* itself was another conspicuous member of the Carboniferous coal forests. (See Fig. 44A.)

Coniferales. Conifers, the pines, spruces, junipers, cypresses, cedars, and their many other living relatives, are familiar to everyone. They arose in the late Carboniferous and reached a climax around the end of the Jurassic and beginning of the Cretaceous. They have since declined in variety and dominance, but everyone knows that they are still extremely abundant in individuals. (See Fig. 44B.)

Gnetales. This is a curious but unimportant little group of living plants including *Ephedra, Gnetum,* and *Welwitschia.* Botanists suppose that the group must be very old, but fossils are known only from the Pliocene.

Subphylum Angiospermae

Angiosperms have true flowers and the seeds are borne covered in ovaries. They include incomparably the greatest number of living land plants. Aside from conifers, almost all the familiar plants of fields, forests, and gardens are angiosperms. Angiosperms are possibly present in the Triassic and definitely in the Jurassic but were still rare in the early Cretaceous. In the late Cretaceous they expanded enormously and became by far the dominant land plants—one of the most remarkable episodes in the history of life. By the end of the Cretaceous and beginning of the Cenozoic most of the main modern groups and even many modern genera were already established. Since then their distributions and associations have changed in interesting ways, but there has been little really fundamental progressive advance. Nor are there any known extinct groups strikingly different from those still living. (See Fig. 44D.)

KINGDOM ANIMALIA

Animals, as here defined, are multicellular organisms with some or much differentiation in forms and functions of different cells. They differ from plants most importantly in being unable to perform photosynthesis and requiring organic food, plants or other animals. Most of them are self-mobile at some stage of life, although many become attached (sessile) at an early age. In correlation with these and other distinctions, tissue and organ differentiation has followed courses very different from those in plants. Development is more closed than in plants, that is, it more definitely tends toward a distinct and fixed adult form. For instance, human babies develop two arms, but oaks develop—how many branches?

Phylum Porifera

Sponges have less extensive cell and organ differentiation than other animals. Typically they have a central cavity and a more or less complex system of tubes, through which water flows, through the body around the cavity. Most sponges have some sort of skeletal stiffening which may consist of scattered spikes or of a complex, coherent network. It may be

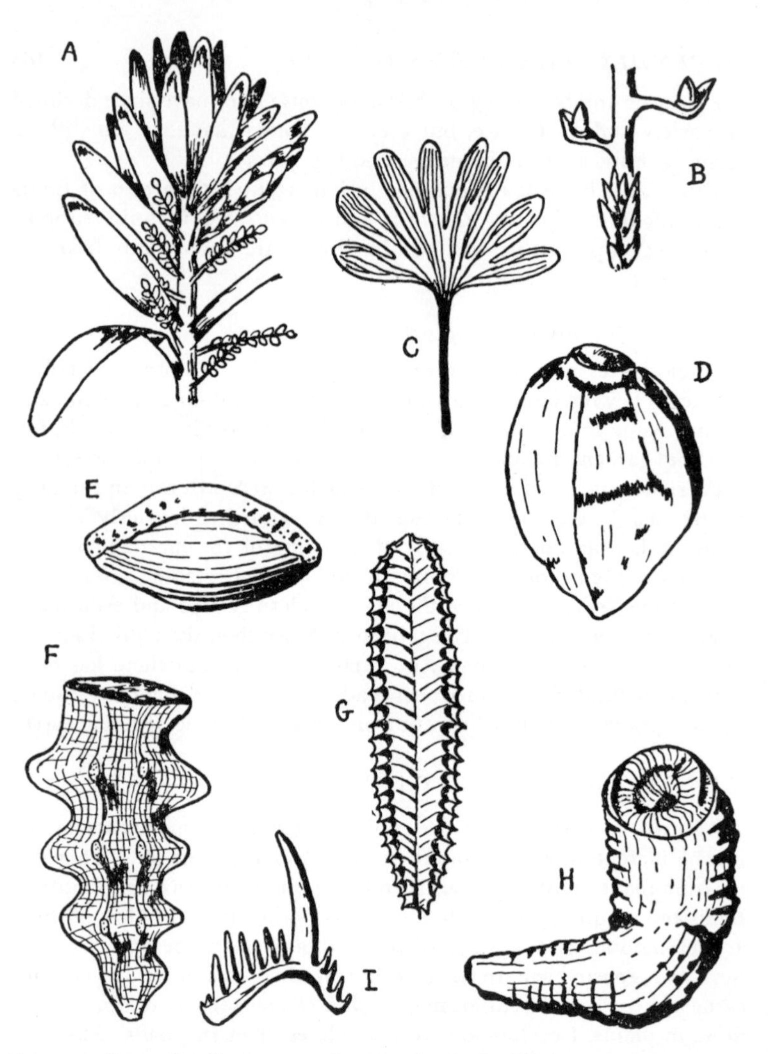

FIG. 44. Some fossil plants and animals. A–D, Phylum Pteropsida. A, Cordaitales, *Cordaites,* Carboniferous, inflorescence and foliage, greatly reduced (after Grand'Eury). B, Coniferales, *Stachyotaxus,* Triassic, seeds and foliage, reduced (after Berry). C, Ginkgoales, leaf of an extinct species of *Ginkgo,* Jurassic, original about 2 inches wide (after Heer). D, Angiospermae, fruit of a palm, *Nipa,* from the London Clay, Eocene, original about 2 inches long (after Bowerbank). E, Protista, Foraminifera, *Triticites,* Carboniferous, original about ¼ inch wide (after Dunbar). F, Phylum Porifera, *Hydnoceras,* internal cast of a siliceous sponge, Devonian, original about 8 inches high (after Hall

composed of calcium carbonate, silica, or spongin, which is the material of a bath sponge. Calcareous and siliceous sponges are rather common fossils from early Cambrian to Recent (see Fig. 44F) but are often difficult to identify and are of minor interest as regards evolutionary change. In the Cambrian there was a peculiar, short-lived group of spongelike animals, Pleospongia or Archaeocyatha, evidently of great importance in early Cambrian seas but soon extinct. Reports of sponge spicules in the pre-Cambrian are not, as yet, worthy of belief.

Phylum Coelenterata

Coelenterates are aquatic animals with a simple central cavity into which there is a single opening, which usually is surrounded by radially arranged tentacles. They are abundant and important throughout most of the fossil record and in the living world. Besides the groups specified below, coelenterates include the jellyfish (Scyphozoa), known as fossils from the Cambrian on but so sporadically as to tell no connected story. The comb jellies (Ctenophora) may be related to coelenterates but are usually considered as a separate phylum. They are unknown as fossils.

Class Hydrozoa

In addition to a number of soft-bodied forms this group includes the hydrocorallines, which have existed since the Triassic, at least, but are not abundant fossils. Some of them, such as *Millepora,* the elk-horn coral, are now important reef-builders. Another sort of encrusting hydrozoan, *Hydractinia,* forms part of the strange fossil *Kerunia,* mentioned in Chapter 5. (See Fig. 24.)

"Class" Stromatoporoidea

This name is applied to an important group of fossils but one that is of quite uncertain status, which is why I have put "class" in quotation marks. The so-called stromatoporoids are calcareous, laminated masses occurring from Cambrian to Permian and forming great reefs in the Silurian and Devonian. Some or all may be ancestral hydrozoans (see above), but the preserved structure is so little characteristic that they may include sponges or members of still some other group.

---

and Clarke). G, "Phylum" Graptolithina, *Tetragraptus,* Ordovician, original about 3 inches long (after Holm). H, Phylum Coelenterata, *Caninia,* a Carboniferous coral, original about 3½ inches along curve (after Missouri Geological Survey). I, phylum uncertain, a conodont, *Euprioniodina,* original about 1/12 inch in width (after Ulrich and Bassler).

Class Anthozoa

Anthozoans include sea anemones, which are soft-bodied and rare as fossils, and corals, which are among the most abundant and interesting of fossils. A stony coral secretes a calcareous base and cup from which the fleshy body or polyp protrudes. Folds in the lower part of the polyp usually occur and in them are formed stony vertical plates, septa, arranged more or less radially inward from the outer wall, the theca. Many corals are solitary, but it is also common for them to form large, connected colonies by budding successively from the original individual. Typical stony corals first appeared in the Ordovician. Other forms occur, but two broad sorts are especially characteristic of the Paleozoic: cornucopia- or hornlike solitary corals, which grew with the small point downward and usually attached to the sea bottom (see Fig. 44H), and tabulate corals. The tabulate corals are colonial, each polyp in a tube with horizontal partitions dividing the current living chamber and those formerly occupied as the coral grew; septa are imperfectly developed or absent. In the Paleozoic most corals with strong septa had them arranged in more or less quadrate patterns and hence are called Tetracoralla. Most of the Mesozoic and Cenozoic corals tend to have a more six-parted arrangement and are called Hexacoralla.

"Phylum" Graptolithina (or Graptozoa)

Graptolites are a puzzling extinct group of small marine animals, common from Cambrian to Silurian and becoming extinct in the early Carboniferous. All were colonial and the fundamental form was an axis along which were arranged from one to four rows of small cups in which the individuals lived. Progressive later forms became branched in various ways and even formed complex networks. Some were attached to the sea bottom, others to floating seaweed, and still others formed their own floats. The skeleton was of chitin and is usually preserved as a dark shiny film, but some are preserved in the round and have been etched from the rock and studied in microscopic detail. In spite of this study, the relationship of these extinct animals is unknown. Many students have referred them confidentially to the Coelenterata or to the Bryozoa (below). Others have recently been just as confident that they are hemichordates (see Chordata, below). It is probable that they are an offshoot of some known phylum, but since we do not really know which one it is less misleading to keep them separate. (See Fig. 44G.)

"Worms" and Conodonts

Soft-bodied animals most of which are summed up as "worms" in the vernacular are extremely common, varied, and ecologically important in the world today. Really they include groups that are fundamentally diverse in anatomy and origin and are properly classified as a number of different phyla. Important among these are: Platyhelminthes, flat worms; Nemertinea, ribbon worms; Nemathelminthes, round worms; Trochelminthes, rotifers and their allies; and Annelida, segmented worms. The annelids, which include the common garden earthworms although most of them are marine, are especially advanced in structure and are very important elements of the animal kingdom. The fossil record is extremely poor for all sorts of worms. For the most part it consists of trails, borings, or tubes that tell little of precise import about the animals that made them. A partial exception is provided by hard parts of annelid jaw mechanisms, which from the Cambrian onward are fairly common as microfossils collectively called scolecodonts.

Another group of microfossils consists of more or less toothlike objects, somewhat resembling scolecodonts, occurring from Ordovician to Permian and called conodonts. Different students have considered them parts of annelids, gastropods, cephalopods, crustaceans, and primitive fishes. The last view seems to be most popular at present, but the plain fact is that no one knows what they are. In spite of this ignorance they are highly characteristic of different ages and useful to geologists on that account. (See Fig. 44I.)

Phylum Bryozoa

Bryozoans, or polyzoans, are tiny, almost always colonial animals superficially resembling some coelenterates but markedly more complex in anatomy. Among other advanced features they have a definitely differentiated alimentary canal with separate anterior and posterior openings, which are, however, situated near each other and not at opposite ends of the animal. Bryozoan colonies secrete characteristic calcareous or sometimes horny supports which are abundant fossils from early Ordovician to Recent. Often they encrust rocks, shells, and the like. In other cases they form lacelike fans and fronds of great delicacy and beauty. (See Fig. 45A.)

Phylum Brachiopoda

Brachiopods or lamp shells are complex marine animals with some resemblances to clams and to bryozoans. The former resemblance is certainly superficial and it is now commonly believed that the latter is also. The adult animal is enclosed in two calcareous or horny shells, or valves, which normally have different shapes, and except in a few extraordinarily aberrant types each valve is symmetrical on the two sides of its midline. Inside the shell are fleshy loops bearing numerous tentacles, and many brachiopods developed elaborate, hard internal supports

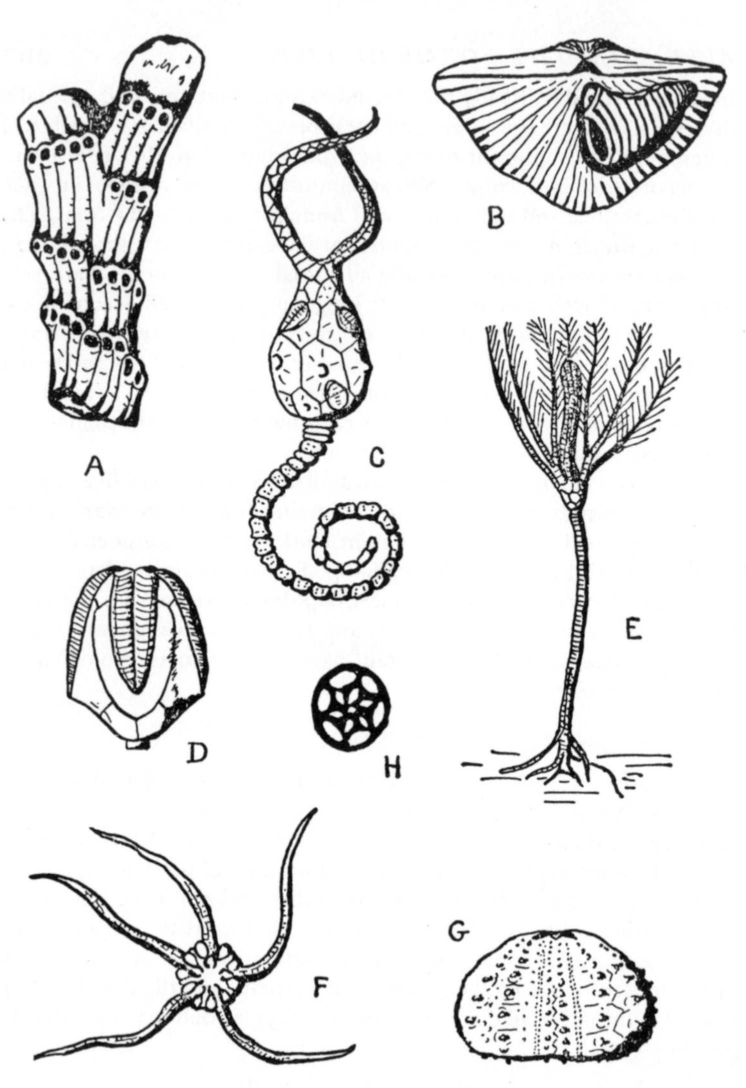

FIG. 45. Some fossil animals. A, Phylum Bryozoa, *Spiropora,* Tertiary, length of original (a fragment of a larger colony) about ¼ inch (after Canu and Bassler). B, Phylum Brachiopoda, *Spirifer,* Carboniferous, with an opening cut on one side to show the spiral internal support (brachidium) for the "arms"; width of original about 2½ inches (compiled diagram, but mainly after Davidson). C–H, Phylum Echinodermata. C, Class Cystoidea, *Pleurocystites,* Ordovician; width of original across calyx about ¾ inch (after Bather). D, Class Blastoidea,

(brachidia) for these loops (lophophores). At the end opposite the opening of the two valves, one valve generally has a notch or hole through which passes a fleshy protuberance, the pedicle, by which adults are usually attached to the sea bottom or some object. The anatomy is rather complex, with fairly well developed digestive, circulatory, and nervous systems. Brachiopods appeared in the early Cambrian and they are extremely abundant in many Paleozoic rocks. Since then they have declined markedly, but many still exist today. In the more primitive forms, Inarticulata, some of which are still living and are remarkable examples of arrested evolution, the two valves were not attached. More numerous and advanced forms, Articulata, developed a strong hinge, sometimes quite complex, between the valves at the pedicle end. (See Fig. 45B, also Figs. 34 and 40.)

Phylum Echinodermata

Echinoderms are also rather complex animals with well-differentiated digestive, circulatory, and nervous systems. They are basically bilaterally symmetrical, but generally this is overlain by a more obvious, five-rayed, radial symmetry. All form calcareous deposits in the body wall, which may be in the form of loose spicules or plates but more often form a complete, hard boxlike protection. Echinoderms are highly varied and are abundant, important fossils. Besides the seven classes listed below specialists recognize several others, of less importance, in the Paleozoic.

Class Cystoidea

Cystoids typically had globular or somewhat flattened bodies encased in a small number of polygonal plates. Some had a small number of arms, also with a system of jointed plates. In most of them the adult was attached by means of a stalk, likewise with plates in columnar arrangement. The group flourished in the Ordovician and Silurian, and became extinct in the Devonian. (See Fig. 45C.)

Pentremites, Carboniferous, calyx only, as usually found; width of original about ½ inch (diagram compiled from various sources). E, Class Crinoidea, *Botryocrinus,* Silurian; height of original about 7 inches (after Bather). F, Class Ophiuroidea, *Ophioderma,* Jurassic; width of original across arms about 7 inches (after Oakley and Muir-Wood). G, Class Echinoidea, *Goniopygus,* Cretaceous; as is usually true of fossil sea urchins, this one has lost its spines; width of original about 1¾ inches (after Clark). H, Class Holothuroidea, a single plate of *Paleochiridota,* Carboniferous; diameter of original about 1/70 inch (after Croneis).

Class Blastoidea

The extinct blastoids had a characteristic small body-box or calyx composed of thirteen principal plates. The stalk was short or absent. Arms were absent. Blastoids lived from Ordovician to Permian and are locally extremely abundant in some early Carboniferous deposits. (See Fig. 45D.)

Class Crinoidea

Crinoids have long been the most varied and abundant stalked echinoderms and are the only ones to survive later than the Paleozoic. They are still common, although less so than in the Paleozoic and now usually not conspicuous because many live in deep water. They appeared in the Cambrian but were rare until the Ordovician, and they became extremely varied in the Silurian through the early Carboniferous. They differ most obviously from cystoids and blastoids in the constant presence of at least five arms, which may branch repeatedly to become very numerous at the tips. Most of them were gregarious and fossil remains all of a single species may be found in enormous swarms. Some limestone beds are made up almost exclusively of separated plates from their columnar stems. A few types became secondarily stemless, and free-living crinoids are sometimes common in the tidal zone of modern coral reefs. The stemmed forms are often called sea lilies because of their (misleading) floral appearance. (See Fig. 45E.)

Class Asteroidea

Asteroids are starfishes, familiar to everyone who has ever frequented a beach. They are free-living echinoderms with a central disk and usually five stout arms. The skeleton consists of many plates, which are not firmly united, so that whole starfishes are uncommon fossils although separate plates and occasional whole animals are known from Ordovician to Recent.

Class Ophiuroidea

The brittle-stars resemble starfish but have the central disk more sharply distinguished from the slender, brittle arms, which are sometimes branched. They are not common fossils but are known from Carboniferous to Recent. (See Fig. 45F.) They arose from a similar but more primitive extinct group, the Auluroidea, Ordovician to Carboniferous, in which the arms were never branched.

Class Echinoidea

Sea urchins or echinoids are encased in globular, heart-shaped, or flattened shells made up of numerous firmly united plates. They are free moving and have numerous spines which are, however, usually detached in fossils. They have existed since the Ordovician and were rather abundant in the Carboniferous but otherwise were not very common as fossils until the Cretaceous. From the Cretaceous to the Recent they have been enormously varied and abundant. (See Fig. 45G.)

Class Holothuroidea

Holothurians, sea cucumbers, are sluggish, leathery, sacklike, free-living echinoderms very different in appearance from other members of this phylum. They have no coherent skeleton, but they secrete small calcareous plates and rods of diverse and characteristic forms. Complete specimens are known from the middle Cambrian, and the calcareous parts are occasional microfossils in later Paleozoic and subsequent rocks. (See Fig. 45H.)

Phylum Mollusca

In spite of the great interest of other groups of invertebrate animals, none compares with the mollusks in abundance as fossils and in value to the paleontologist. Most of them have hard shells which fossilize readily and are fully characteristic and indicative of their relationships. They have been among the most varied of animals ever since the Cambrian. They also have representatives in a wide range of environments, in the sea, from its abyssal depths to tidal flats, in lakes and streams, and on dry land from shore to mountain tops. Anatomically they are more complex than any animals except arthropods and vertebrates (see below), with, in advanced forms, at least, a complex circulatory system including a chambered heart, a digestive system with differentiated liver, kidneys, and a nervous and sensory system that in some high types includes eyes functionally quite like ours and ganglia so complex as to merit designation as brains.

Class Pelecypoda

Pelecypods or lamellibranchs are the bivalves, many of which are familiar to everyone as seen on the beach or on the table. They include clams, oysters, mussels, cockles or scallops, and literally thousands of other forms. The two shells or valves differ from those of brachiopods (above) in that each one is asymmetrical, but typically the two are nearly symmetrical with respect to each other—that is, one is approximately a

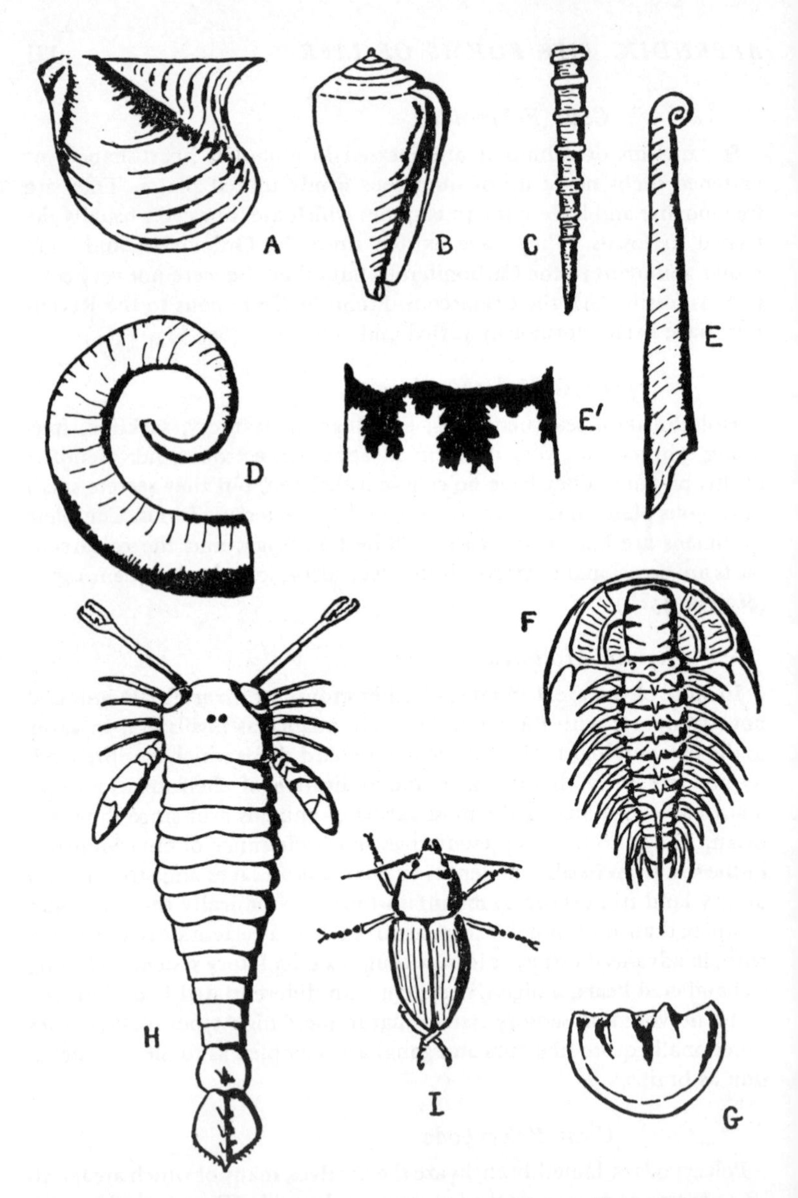

FIG. 46. Some fossil animals. A, Phylum Mollusca, Class Pelecypoda, *Limoptera,* Devonian; width of original about 1¾ inches (after Hall). B, Phylum Mollusca, class Gastropoda, *Conus,* Tertiary; height of original about ¾ inch (after Heilprin). C, Phylum Mollusca, Class Gastropoda (?), *Tentaculites,* Silurian; length of original about ½ inch

mirror image of the other. (But in a number of specialized groups of pelecypods one valve did become quite different from the other.) Some pelecypods become attached, as do oysters, but most are free, moving and burrowing with a somewhat hatchet-shaped foot (clams) or swimming by clapping the two shells (scallops). The great majority are and have been clamlike, but some very bizarre divergent types have occurred; to name only one: the Cretaceous *Toucasia* in which each of the two valves was strongly spiral and the whole animal looked like two snails pressed together. Pelecypods occur in the Cambrian but were then very rare. From Ordovician to Recent they are among the most abundant marine fossils and are also common in some fresh-water deposits. (See Fig. 46A, also Figs. 11 and 38.)

Class Gastropoda

Gastropods, or snails in a very broad sense, are even more numerous than pelecypods today and are equally or more common as fossils from early Cambrian to Recent. Some of them are bare, like the garden slugs or marine nudibranchs, but most of them have a single shell, the opening of which is, however, often covered by a second small calcareous or horny shell, the operculum, when the animal withdraws into its shell. The shell may be saucer-shaped, conical, or coiled in a plane, but it is usually a more or less elevated spiral. Even within the generally spiral pattern, form and ornamentation are extremely varied and characteristic, so that the abundant fossils are excellent material for evolutionary studies. Almost all gastropods are free-living, and some are fiercely

---

(after Hall). D, Phylum Mollusca, Class Cephalopoda, Nautiloidea, *Ryticeras,* Devonian; original about 5 inches greatest distance across whole shell (after Hall). E, Phylum Mollusca, Class Cephalopoda, Ammonoidea, *Baculites,* Cretaceous, a very young shell, original length about 1/3 inch; E′ same, part of an adult shell showing suture line formed by the fluted internal partition, original width about 1 3/4 inches (both E and E′ after Reeside). F, Phylum Arthropoda, Class Trilobita, *Zacanthoides,* Cambrian; original total length, including spine, about 1 1/2 inches (after Walcott). G, Phylum Arthropoda, Class Crustacea, *Kyamodes,* a Silurian ostracod shell, width of original a little over 1/16 inch (after Ulrich and Bassler). H, Phylum Arthropoda, Class Arachnida, Eurypterida, *Pterygotus,* Silurian, length of body of original about 5 1/2 feet (after Clarke and Ruedemann). I, Phylum Arthropoda, Class Insecta, *Tauredon,* a Jurassic beetle, length of original about 1 inch (after Handlirsch).

carnivorous, preying especially on other mollusks, but the majority is herbivorous. Like pelecypods, they are common in all aquatic environments, and they also include many land-living forms that have a lung-like sack for breathing air. (See Fig. 46B, also Fig. 4C.)

Among the most peculiar gastropods are the pteropods, pelagic forms of the open sea with conical or pyramidal shells that contribute largely to the oozes of the ocean floor. There are some extremely abundant Paleozoic fossils, notably one called *Tentaculites,* that resemble pteropod shells and have often been placed in that group, but recently students question the relationship. *Tentaculites* may not even be a mollusk. (See Fig. 46C.)

Class Amphineura

Naked, wormlike amphineurans seem to be very primitive and have been hailed as survivors of a group ancestral to all mollusks, but they are unknown as fossils and they may merely represent late degeneration. The familiar chitons, with a series of eight plates down the back, occur as fossils from Ordovician onward; the ancient forms are quite like those still living.

Class Scaphopoda

Scaphopods are marine mollusks, many of which have elongated, tubular shells, open at both ends and usually somewhat curved. Scaphopods seem always to have been a rather rare and conservative group. The earliest forms, Silurian, differ little from the living genus *Dentalium.*

Class Cephalopoda

By human criteria, at least, cephalopods represent the culmination of one of the two most progressive lines of invertebrate development (the other is that of the insects). They are also common fossils of high value and interest. All are marine, free-living carnivores with tentacles surrounding the mouth. Sensory and nervous systems are as complex as in any invertebrates. Nautiloids, with straight or variously coiled external shells divided into chambers by relatively simple partitions, appeared sparsely—some students say dubiously—in the Cambrian and were very common in the Ordovician and Devonian. (See Fig. 46D.) Since then they have had a complex and interesting history but have declined to a single living genus, the chambered nautilus. The related ammonoids typically had plane-coiled shells rather similar to that of a nautilus but with more complex partitions. Later forms developed a great variety of shell shapes. They appeared in the Silurian and are very abundant, extremely interesting and valuable fossils in the later Paleozoic and throughout the Mesozoic, but they became extinct by the end of the Cretaceous. (See Fig. 46E, also Figs. 10 and 21.)

Another group, the coleoids or dibranchiates, is common today, including squids, octopuses, and their allies, but has a much less extensive fossil record. They have no shells or have reduced and mainly internal shells. Belemnites, forerunners of the cuttlefishes, are fairly (in some particular localities, very) common fossils in the Mesozoic. All that is usually preserved is a dense cylinder, pointed at one end and hollow at the other. Our ancestors long knew these strange fossils (see Fig. 2, Chapter 2) and thought they were literally thunderbolts, among other curious and fantastic interpretations.

Phylum Arthropoda

Jointed animals, usually with external, chitinous skeletons, are another culmination of invertebrate evolution and are the dominant organisms on earth today in terms of abundance and variety. Specialists subdivide them into numerous groups, but only four are of great importance in the economy of nature and have extensive fossil records.

Class Trilobita

The extinct (Cambrian to Permian) trilobites are related and perhaps through early forms ancestral to the crustaceans, but most specialists now place them in a separate class. Head and body were divided into three longitudinal lobes (hence the name) and usually also into numerous transverse segments, each with a complex, jointed appendage. Trilobites are abundant Paleozoic marine fossils and their history can be followed in great detail. (See Fig. 46F.)

Class Crustacea

This class is extremely varied and has been split into various separate groups by some students. It includes the familiar crabs, lobsters, shrimps, crayfish, and their many allies, as well as barnacles and several less widely familiar groups. True crustaceans, as opposed to trilobites, also appeared in the early Cambrian, but the group characterized by crabs, lobsters, and the like (Decapoda) did not develop until the Triassic. (See Fig. 24.) Most abundant as fossils and useful in dating and correlating rocks are the ostracods, tiny crustaceans that secrete two symmetrical shells. They occur from early Ordovician to Recent. (See Fig. 46G.)

Class Arachnida

Familiar living arachnids are scorpions, spiders, mites, ticks, and the horseshoe "crab" *(Limulus)*, which is not a crustacean but the little-changed, relict survivor of an ancient group of arachnids. The class appears in the early Cambrian and has a complex history. An extinct group of special interest is that of the Eurypterida, Ordovician to Permian, somewhat scorpion-like aquatic forms some of which reached enormous sizes. *Pterygotus,* the "seraphim" of Scottish quarrymen, attained a length of well over six feet. (See Fig. 46H.)

Class Insecta

The dominance of insects in the world today is obvious to all, and their general characteristics and main varieties are familiar. They appeared in the Devonian, and large dragonfly- and cockroach-like types were especially striking in the Carboniferous. A very large number of fossil insects is known, but the record is spotty and does not as yet adequately reflect their true abundance and details of their extremely complex history. Small, flying animals, especially, are preserved as fossils only under rather unusual conditions. (See Fig. 46I, also Figs. 32 and 33.)

Phylum Chordata

Chordates are generally bilaterally symmetrical, free-living animals with relatively very high differentiation of organs, notably the nervous system, and with a stiffening (notochord or backbone) down the back. There are living today several sorts of presumably primitive chordates without bony skeletons: hemichordates (tongue or acorn worms), tunicates (including the ascidians or sea squirts), and cephalochordates (lancelets, amphioxus). They cast some—rather feeble—light on the origin of the phylum, but are doubtfully or not known as fossils. As mentioned above, there is a theory that the abundant Paleozoic graptolites are hemichordates, but this is no more than a possibility and would not help much in understanding the rise of the chordates even if true.

The more important chordates are the vertebrates, with well-developed bony or, less commonly, cartilaginous internal skeletons, including a jointed backbone. Most modern classifications divide them into eight classes, each of which will be briefly reviewed.

Class Agnatha

Agnathans are aquatic, fishlike forms in which true, movable jaws have not developed. They appeared in the late Ordovician and were briefly the dominant vertebrates in the late Silurian. Among the early forms were ostracoderms or Osteostraci, late Silurian to late Devonian, with rigid, flattened head shields (see Fig. 35). Also with head shields but less or not flattened in form were the pteraspids, late Ordovician to late Devonian (see Fig. 47A). The anaspids, late Silurian to late Devonian, were most fishlike in appearance and characteristically had the head covered by many small plates or scales rather than by a shield. (See Fig. 12E, F.) After the Devonian there is an enormous, complete gap in the fossil record, but the living hagfishes and lampreys undoubtedly are survivors of the Devonian agnathans.

Class Placodermi

The placoderms, late Silurian to Permian, are a rather miscellaneous group of fishlike, aquatic vertebrates in which movable jaws had developed but the skeletal arches behind the jaws were still unspecialized. They include six or more separate groups which may not really have been very closely related. The arthrodires, some of which reached large, sharklike sizes, had heavy head armor and large, powerful jaws. The smaller antiarchs, including *Bothriolepis,* also had heavy armor, but jaws so small that they were at first overlooked and the animals were confused with ostracoderms. The numerous acanthodians, longest-lived of placoderms, looked superficially like small, spiny sharks. (See Fig. 47B, also Figs. 12A, and 35.) No placoderms survived as such after the Paleozoic, but early placoderms gave rise to the two classes of true fishes and live on in them.

Class Chondrichthyes

The true fishes of today, with jaws and with specialized arches, especially the hyoid, posterior to the jaws, belong to two grand divisions so distinctive that their students now place them in separate classes. Both appeared in the middle Devonian, or perhaps at the end of early Devonian, and both are common today. Both have relatively fine, continuous fossil records. The Chondrichthyes are the sharks, skates, rays, chimaeras, and their many living and extinct allies. The later fossils and all the living forms are characterized by having cartilaginous rather than bony skeletons, but early representatives had extensive bone. (See Fig. 47C, also Fig. 12A, B.)

Class Osteichthyes

Arising at the same time as the Chondrichthyes, the true bony fishes early outstripped the sharks and their relatives and have long been by far the dominant fishes. The more primitive bony fishes (chondrosteans), dominant in the Paleozoic, are represented now by a few relicts, including the sturgeons, which are no longer bony. Through complex latest Paleozoic and Triassic groups arose the teleost fishes, which have expanded enormously since the early Jurassic and now include by far the greatest number of both marine and fresh-water fishes.

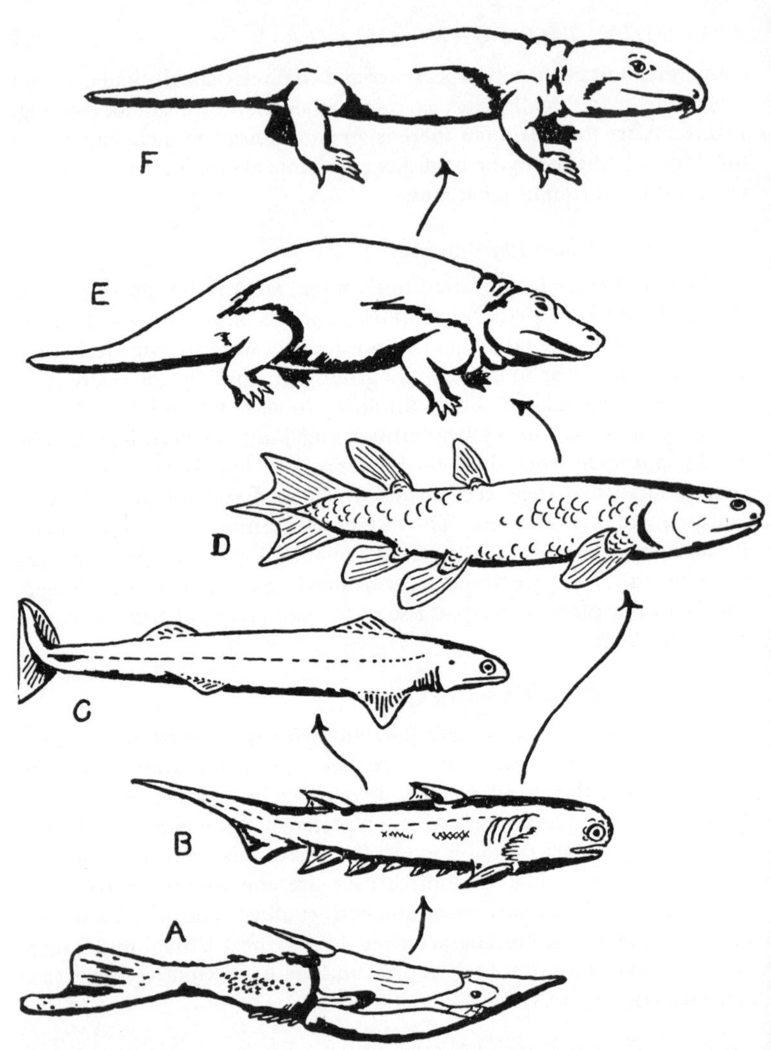

FIG. 47. Restorations of some fossil vertebrates. A, class Agnatha, *Pteraspis,* Devonian, length of original about 9 inches (after White). B, Class Placodermi, *Climatius* (an acanthodian), Devonian, length of original about 3 inches (restored from a partial skeletal reconstruction by Watson). C, Class Chondrichthyes, *Cladoselache,* Carboniferous, length of original about 3 feet (after Dean). D, Class Osteichthyes, *Eusthenopteron,* a crossopterygian, Devonian, length of original about 2½ feet (restored from a skeletal reconstruction by Gregory and Raven). E, Class Amphibia, *Eryops,* a labyrinthodont, Permian, length

Quite a different development of Osteichthyes is represented by the crossopterygians and the dipnoans or lungfishes. Lungfishes were common in the Devonian but declined thereafter and are now represented by relicts in Australia, South America, and Africa. Crossopterygians also flourished in the Devonian and then gave rise to the amphibians. It was long supposed that crossopterygians, as such, became extinct at the end of the Cretaceous, but at least one living genus, *Latimeria,* was recently found alive off the east coast of Africa. (See Fig. 47D.)

Class Amphibia

Amphibians arose from crossopterygian fishes in the late Devonian, from which almost perfectly transitional forms are known. Most of the ancient, Devonian to Triassic, amphibians are grouped together as labyrinthodonts. They were extremely varied, but most of them were larger than recent amphibians, with heavy, flattened skulls, four usually feeble legs of about equal length, and stubby tails. (See Fig. 47E, also Figs. 9, 35.) The frogs, toads, and their allies (Anura), with no tails in the adults and with elongated, leaping posterior legs, had somewhat questionable allies as early as the late Carboniferous but are first definitely recognizable in the late Jurassic. They are now much the most common surviving amphibians. The salamanders and their relatives (Urodela) resemble lizards, for which they are usually mistaken by the uninformed. They are unknown before the early Cretaceous and their origin is a disputed mystery. There are several other small groups of amphibians of less interest for this brief summary.

Class Reptilia

The reptiles, distinguished from amphibians by elimination of an early aquatic stage in development, arose from labyrinthodonts in the late Carboniferous and were the dominant land animals of the Mesozoic, often called the Age of Reptiles. (In fact, they were equally dominant in the Permian.) Conservative classifications recognize at least fifteen major groups (orders) of reptiles and some have up to twice that number. Only a few major sorts can be briefly reviewed here.

Cotylosaurs. These were the primitive reptiles, still rather labyrintho-

---

of original about 5 feet (restored from a skeleton in the American Museum). F, Class Reptilia, *Labidosaurus,* a cotylosaur, Permian, length of original about 2 feet (restored from a skeletal reconstruction by Williston). The arrows indicate descent of the various classes, not of the particular examples shown.

dont-like but with generally stronger legs and better adapted to dry land. They are known from late Carboniferous to late Triassic, and all other reptiles are believed to have originated from them. (See Fig. 47F.)

Turtles, etc. Tortoises, turtles, and the other shelled reptiles constituting the Chelonia, a group familiar to everyone today, have a doubtfully claimed ancestor in the middle Permian and are well represented by fossils from the late Triassic onward.

Ichthyosaurs. The ichthyosaurs were completely aquatic, viviparous marine reptiles similar in appearance and habits to dolphins, which are mammals. They have a good fossil record from middle Triassic to late Cretaceous, when they became extinct. (See Figs. 41, 42D.)

Plesiosaurs. The plesiosaurs have the same time range as ichthyosaurs and were also marine but were quite different in appearance. They had a plump, turtle-like but not shelled body, four large flippers, and usually a rather long, sometimes extraordinarily long, neck. Someone has remarked that they must have looked somewhat like a turtle with a snake threaded through the shell.

Rhynchocephalians. This has never been a very common group, but it is worthy of mention if only for the unique survivor, *Sphenodon,* of New Zealand, a lizard-like animal that has hardly changed since the Jurassic. Somewhat larger allies were characteristic of the Triassic.

Lizards and snakes. These two familiar groups are closely allied and belong in the same order (Squamata). Lizards are more primitive and appeared in the Jurassic but were not particularly abundant until the Cenozoic. Snakes probably arose from limbless burrowing lizards, although snakes now have a great variety of other habits and habitats. They date from the early Cretaceous but were also rare until well into the Cenozoic.

Crocodiles. Crocodiles arose at the end of the Triassic and ecologically replaced an older similar group, the phytosaurs, which flourished in the Triassic but were confined to that period. Crocodiles of many different sorts, including some completely marine, were abundant in the Jurassic and Cretaceous. Forms rather like those now living, including alligators, gavials, etc., are common throughout the Cenozoic.

Pterodactyls. These are the batlike flying reptiles of the Mesozoic, early Jurassic to late Cretaceous.

Dinosaurs. The animals popularly called dinosaurs, lords of the Mesozoic, belong to two quite distinct groups, the orders Saurischia and Ornithischia. In the first order are the theropods, bipedal, mostly carnivorous, late Triassic to late Cretaceous; and the sauropods, largest of all dinosaurs, quadrupedal, amphibious, herbivorous, with small heads and long necks and tails, early Jurassic to late Cretaceous, but especially characteristic of late Jurassic. The Ornithischia, all of which were herbivorous, include four quite different sorts of dinosaurs: bipedal, unarmored forms, such as *Iguanodon,* the camptosaurs, and the hadrosaurs or duckbills, middle Jurassic to late Cretaceous; the stegosaurs, quadrupeds with plates down the back and spikes on the tail, early Jurassic to early Cretaceous; the ankylosaurs, quadrupeds with the body encased in bony armor, early to late Cretaceous; and the ceratopsians, quadrupeds with very large heads and a neck frill projecting backward from the skull, usually also with horns, confined to the late Cretaceous. (See Figs. 16 and 17.)

Pelycosaurs. These were forerunners of the mammal-like reptiles. They lived from late Carboniferous to late Permian, doubtfully into earliest Triassic. Some of them had enormously elongated spines down the back. (See Fig. 14.)

Mammal-like reptiles. The first definitely mammal-like reptiles (Order Therapsida) appeared in the middle Permian and the last in the middle Jurassic. In the meantime they had given rise to the mammals, and they include our own ancestors of Permian and Triassic times. The group was extremely varied, especially in South Africa from which they are best known, although in the Triassic, at least, they occurred all over the world. (See Fig. 48C.)

Class Aves

The oldest known bird, still hardly more than a feathered reptile, is *Archaeopteryx* of the middle Jurassic. (See Fig. 48B.) Some late Cretaceous birds still had teeth but were otherwise highly specialized and completely birdlike. Birds expanded and diversified greatly in the latest Cretaceous and early Cenozoic. By the Miocene almost all the main recent groups were in existence. In general their fossil record is poor. Of special interest are some very large, extinct, flightless birds: the moas of New Zealand, *Aepyornis* in Madagascar, the phororhacids and brontornids of South America, and the diatrymids of North America and Europe.

Class Mammalia

Mammals differ most importantly from reptiles in physiological characters, especially in bearing living young (viviparity), giving milk, and maintaining fairly even body temperatures (homothermy). These characters are not directly determinable in fossils and some, perhaps all, of them probably arose in animals classified as mammal-like reptiles. Among fossils an arbitrary division is made by calling an animal a reptile if the tooth-bearing bone of the lower jaw (the dentary) does

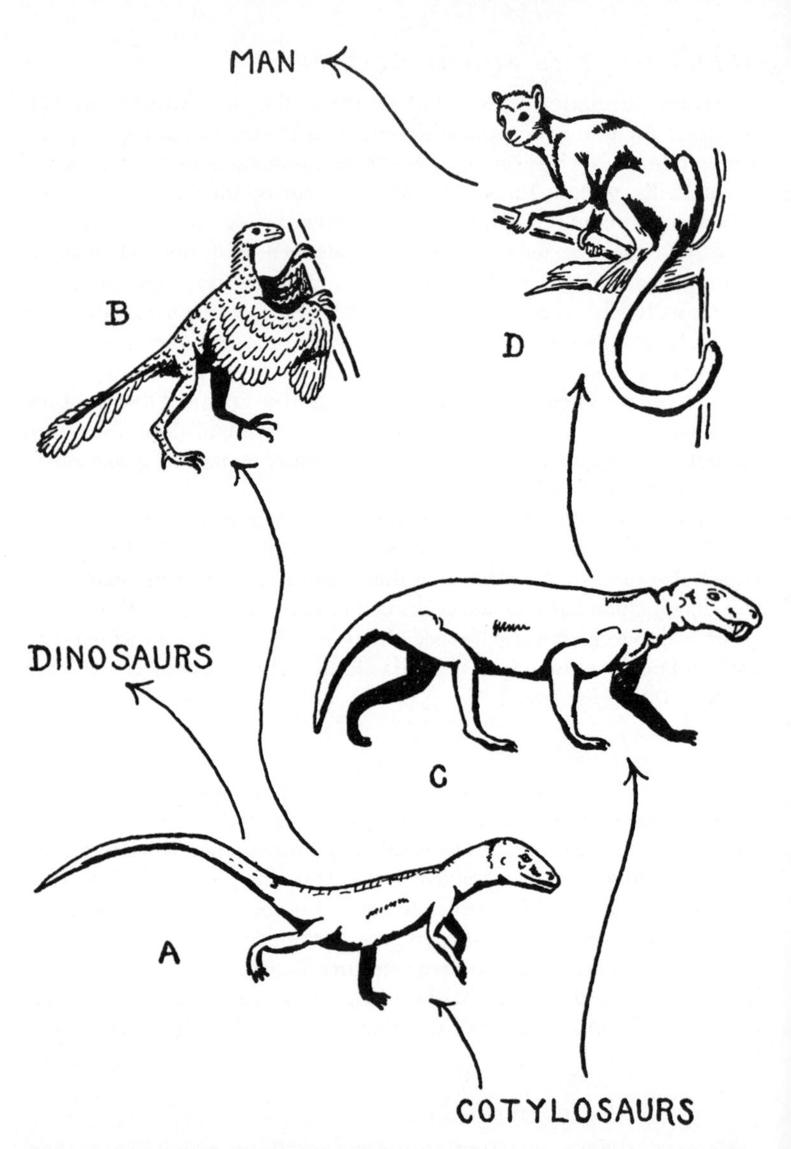

FIG. 48. Restorations of some fossil vertebrates (a continuation of Fig. 47). A, Class Reptilia, *Ornithosuchus,* a pseudosuchian (the general group from which dinosaurs, birds, crocodiles, lizards, etc. were derived), Triassic, length of original about 3 feet (restored from a skeletal reconstruction by von Huene; the pose, suggested by Heilmann, may be misleading to the extent that the animal probably walked four-footed unless startled). B, Class Aves, *Archaeopteryx,* Jurassic, length

not articulate directly with the skull, and calling the fossil a mammal if it does.

Conservative modern classification recognizes 33 orders of mammals, of which 15 are extinct and most have good fossil records. Summary must be abbreviated and some minor groups omitted.

Monotremes. The egg-laying "mammals" of Australia have no useful fossil record but must be mentioned because of their special interest. It was long thought that they represented survivors of a stage of evolution ancestral to all other mammals. This is, however, extremely improbable. It is more likely that they are survivors of some line of mammal-like reptiles not ancestral to true or other mammals. Although the lower jaw bone does articulate directly with the skull and they do give milk, the monotremes might be classified as therapsid reptiles rather than as mammals. But one high authority, W. K. Gregory, thinks they are degenerate marsupials.

Multituberculates. This long-lived, ancient group probably represents a separate development from mammal-like reptiles, distinct from other, or true, mammals. The animals were small and rodent-like in habits. They appeared in the late Jurassic and survived until early Eocene when true rodents had appeared and took over their ecological role.

Pantotheres. Mammals are possibly represented by single teeth in latest Triassic or earliest Jurassic, but the first surely established mammals (by definition) occur in the middle and late Jurassic. There are several groups at that time, among which the most important are the pantotheres because they seem to represent the ancestry of almost all later mammals. They are tiny creatures, known from quite fragmentary specimens, shrewlike in size and in some tooth characters.

Marsupials. These are the mammals in which the young are born in what seems to us a premature condition and usually complete their development in a pouch on the mother's abdomen. They are known from the late Cretaceous to Recent. As everyone knows, they are highly varied in Australia and include most of the mammals of that continent, but their fossil record there is still extremely poor, probably from lack of sufficient searching. They are also common now (mostly opossums, Di-

---

of original about 1½ feet (after Gregory, based on skeletal reconstruction by Heilmann). C, Class Reptilia, *Lycaenops,* a mammal-like reptile (therapsid), Permian, length of original about 4 feet (restored from a skeletal reconstruction by Colbert). D, Class Mammalia, *Notharctus,* a primitive primate (lemuroid), Eocene, length of head of original about 3 inches (after Gregory). The arrows indicate descent of the general groups represented, not necessarily of these particular examples.

delphidae) in North and South America and formerly constituted a larger and more varied part of the fauna of the latter continent. Opossums (true, not Australian) also occurred in Europe in early and middle Cenozoic.

Insectivores. All the later, nonmarsupial mammals are called placentals, from the nourishment of the foetus through a placenta and its birth in a relatively advanced stage. Among placentals the Order Insectivora, not all members of which are insectivorous in food habits, is a miscellaneous group including the oldest, most primitive forms and numerous others that do not happen to have progressed decisively enough to be given a different name. Insectivores appear in the late Cretaceous and are rather numerous thereafter. Living examples are shrews, moles, and hedgehogs. (See Fig. 4B.)

Bats. These flying mammals have a poor fossil record, but the bat wing is known to have been fully developed by middle Eocene, at latest. (See Fig. 32.)

Primates. This order, to which we ourselves belong, arose from the Insectivora, from which various early and primitive primates cannot be really clearly distinguished. Definite but still primitive primates, prosimians, appear in the middle Paleocene and were common in Europe and North America during the Eocene. (See Fig. 48D.) In the Oligocene two distinct groups of monkeys had arisen in, respectively, South America and the Old World, and forerunners of the great apes were also present. Great apes were more varied than now in the Miocene. Man arose only toward the end of the Cenozoic and definitely human fossils are not known until the Pleistocene.

Edentates. Edentates, which may lack teeth as the name implies, but which more typically have simplified, enamelless teeth, first appear in the record in the late Paleocene of North America. They developed greatly in South America and are characteristic of that continent. Extinct groups are the large, armored glyptodonts (late Eocene to Pleistocene) and the ground sloths (late Eocene or early Oligocene to Pleistocene), both of which also invaded North America late in the Cenozoic. Living are the armadillos (since late Paleocene), anteaters (since early Miocene), and tree sloths, not surely known as fossils but apparently derived from Miocene ground sloths.

Lagomorphs and rodents. Although commonly called rodents, the hares, rabbits, and pikas belong to quite a distinct group, Lagomorpha, known in the late Paleocene but common only in Oligocene and later times. True rodents also appeared sparsely in the late Paleocene and began to expand in the Eocene. They have become much the most varied and widespread group of mammals and are abundant fossils, occurring in practically all known land faunas from early Oligocene onward. Their history is extremely complex but is beginning to be well known.

Cetaceans. The whales, dolphins, porpoises, and their allies are completely aquatic mammals. Large but primitive forms, archaeocetes, appeared already entirely aquatic in the middle Eocene and survived to early Miocene. Next arose the toothed whales, odontocetes, late Eocene to Recent, including sperm whales, dolphins, and porpoises, and finally the whale-bone whales, mysticetes, middle Oligocene to Recent. (See Figs. 41, 42E.)

Carnivores. The carnivores were of common origin with the ungulates and barely separable from them in the early Paleocene. Archaic forms, creodonts, dominated Paleocene faunas, and modern types began to appear around the beginning of the Oligocene: dogs, cats, raccoons, weasels, and civets. Bears arose from dogs in the Miocene and hyenas from civets in the Pliocene. The long and intricate history is documented by many fossils. (See Fig. 4A.)

Ungulates. The (mostly) hoofed herbivores are a very complex group of animals not all of closely similar origin and now classified in fourteen or more different orders. The majority of these may have been separately derived from allies of the most primitive Paleocene and Eocene ungulates called Condylarthra. Among particularly important other and later groups are: Notoungulata, Paleocene to Pleistocene in South America, where they were the old native ungulates and to which they were almost confined; Proboscidea, the mastodonts, mammoths (which are merely extinct species of elephants), and elephants, late Eocene to Recent and eventually spread over most of the world except Australia (see Fig. 13); Perissodactyla, horses, tapirs, rhinoceroses, and a host of extinct forms, early Eocene to Recent (see Figs. 3A, 15, 39); and Artiodactyla, pigs, hippopotamuses, deer, giraffes, cattle, sheep, antelopes, and an even larger host of relatives, early Eocene to Recent (see Fig. 37). The sea cows, Sirenia, including the living dugongs and manatees, are apparently of ungulate origin although they early became fully aquatic. They are fairly common fossils in some deposits since the Eocene. There are several other striking but relatively unvaried and short-lived extinct orders of ungulates, among them the Pantodonta, middle Paleocene to middle Oligocene, including Coryphodon (met in Chapter 1), and the Dinocerata or uintatheres, great horned beasts known from late Paleocene to late Eocene in North America and Asia.

Suggestions for Further Reading

ANDREWS, H. N., JR. 1947. Ancient plants and the world they lived in. Comstock, Ithaca. Popular paleobotany.

ARNOLD, C. A. 1947. An introduction to paleobotany. McGraw-Hill, New York. College text; requires some knowledge of botany.

BUCHSBAUM, R. 1948. Animals without backbones. An introduction to the invertebrates. 2d ed. Univ. Chicago Press, Chicago. Easy text on recent invertebrates; richly illustrated.

CAMP, C. L., and G. D. HANNA. 1937. Methods in paleontology. Univ. California Press, Berkeley. Collection and preparation of fossils.

CARTER, G. S. 1951. Animal evolution. A study of recent views of its causes. Sidgwick and Jackson, London. Well-rounded, modern, semitechnical summary.

CLARK, W. E. LEGROS. 1950. History of the primates. An introduction to the study of fossil man. 2d ed. British Museum, London. Brief, popular, authoritative.

COLBERT, E. H. 1951. The dinosaur book. 2d ed. McGraw-Hill, New York. Popular account of all kinds of fossil reptiles.

DOBZHANSKY, TH. 1951. Genetics and the origin of species. 3d ed. Columbia Univ. Press, New York. Genetical basis of evolutionary theory; technical, requires knowledge of general genetics.

DUNBAR, C. O. 1949. Historical geology. Wiley, New York. Beginning college text; much on fossils.

GEORGE, T. N. 1951. Evolution in outline. Thrift Books, London. Brief, modern, popular, emphasis on paleontology.

GILLISPIE, C. C. 1951. Genesis and geology. A study in the relations of scientific thought, natural theology, and social opinion in Great Britain, 1790–1850. Harvard Univ. Press, Cambridge. History of the period formative for modern geology and paleontology.

GILLULY, J., A. C. WATERS, and A. O. WOODFORD. 1951. Principles of geology. Freeman, San Francisco. Geological processes rather than history, but essential basis for understanding of earth history.

GLAESSNER, M. F. 1945. Principles of micropaleontology. Melbourne Univ. Press, Melbourne. Advanced textbook.

GOLDRING, W. 1950. Handbook of paleontology for beginners and amateurs. Part 1, The fossils. New York State Museum, Albany. Descriptive; emphasis on New York fossils.

HESSE, R., W. C. ALLEE, and K. P. SCHMIDT. 1951. Ecological animal geography. 2d ed. Wiley, New York. Moderately technical review of ecological and zoogeographical principles as applied to recent animals.

HUXLEY, J. 1942. Evolution. The modern synthesis. Harper, New York. Semitechnical; little paleontology but otherwise thorough.

KRUMBEIN, W. C., and L. L. SLOSS. 1951. Stratigraphy and sedimentation. Freeman, San Francisco. University text, including paleontological stratigraphy; requires some knowledge of general geology.

MAYR, E. 1942. Systematics and the origin of species. Columbia Univ. Press, New York. Population systematics and evolution; technical.

MOORE, R. C. 1949. Introduction to historical geology. McGraw-Hill, New York. Beginning college text; much on fossils.

MOORE, R. C., C. G. LALICKER, and A. G. FISHER. 1952. Invertebrate fossils. McGraw-Hill, New York. Detailed descriptive text.

ROMER, A. S. 1941. Man and the vertebrates. 3d ed. Univ. Chicago Press, Chicago. Introductory vertebrate zoology; much on fossils.

———. 1945. Vertebrate paleontology. 2d ed. Univ. Chicago Press, Chicago. Advanced text.

SHIMER, H. W., and R. R. SHROCK. 1944. Index fossils of North America. Wiley, New York. Technical aid to identification of American fossil invertebrates.

SIMPSON, G. G. 1949. The meaning of evolution. Yale Univ. Press, New Haven. Popular summary of the fossil record, evolutionary theory, and their ethical bearings.

———. 1951. Horses. Oxford Univ. Press, New York. Popular account of recent and fossil horses and their evolution.

———. 1953. The major features of evolution. Columbia Univ. Press, New York. Technical treatise on evolution and the fossil record.

SINNOTT, E. W., L. C. DUNN, and TH. DOBZHANSKY. 1950. Principles of genetics. McGraw-Hill, New York. Introductory college text; ample basis for genetical aspects of evolution.

STORER, T. I. 1951. General zoology. 2d ed. McGraw-Hill, New York. College text; adequate review of recent animals.

SWINNERTON, H. H. 1947. Outlines of paleontology. 3d ed. Arnold, London. Beginning text.

TRANSEAU, E. N., H. C. SAMPSON, and L. H. TIFFANY. 1940. Textbook of botany. Harper, New York. Beginning text; suitable background for more advanced paleobotany.

YOUNG, J. Z. 1950. The life of vertebrates. Clarendon, Oxford. Semitechnical; summaries of fossil relatives of major groups.

ZEUNER, F. E. 1946. Dating the past. Methuen, London. Technical treatise on determination of geological time.

ZITTEL, K. A. 1901. History of geology and paleontology to the end of the nineteenth century. Scribner, New York. Dry but informative history.

Index

Page numbers marked with an asterisk (*) indicate an illustration